Hex Strategy

Hex Strategy:
Making the Right Connections

Cameron Browne
Cyberite Pty Ltd.

CRC Press
Taylor & Francis Group
Boca Raton London New York

CRC Press is an imprint of the
Taylor & Francis Group, an **informa** business

CRC Press
Taylor & Francis Group
6000 Broken Sound Parkway NW, Suite 300
Boca Raton, FL 33487-2742

Visit the Taylor & Francis Web site at
http://www.taylorandfrancis.com

and the CRC Press Web site at
http://www.crcpress.com

Dedicated to Helen

For remaining oblivious
to the whole thing.

Table of Contents

Preface xi

1 Introduction 1
    1.1 The Game of Hex 1
    1.2 Rules of Play 2
    1.3 History 3
    1.4 Nature of the Game 4
    1.5 The Shannon Switching Game 8
    1.6 Current Literature 13
    1.7 Board Representations 17
    Summary 23

2 Adjacency and Connectivity 25
    2.1 Coordinate Systems and Adjacency 25
    2.2 Chains 27
    2.3 Connectivity 28
    Summary 42

3 Strategy I: Basic 43
    3.1 Structural Development 43
    3.2 Positional Play 46
    3.3 General Strategies 49
    3.4 Applying Basic Strategy 51
    Summary 52

4 Groups, Steps, and Paths 53
    4.1 Groups 53
    4.2 Steps 56
    4.3 Paths 59

    4.4 Groups from Paths 65
    Summary 67

5   Templates 69
    5.1 Connection Templates 69
    5.2 Template Intrusions 77
    5.3 Multi-Piece Edge Templates 78
    Summary 79

6   Strategy II: Intermediate 81
    6.1 Expand With Templates 81
    6.2 Momentum 86
    6.3 Forcing Moves 87
    6.4 Home Area 92
    6.5 Edge Awareness 94
    6.6 Loose Connections 97
    6.7 Reduce the Opponent's Alternatives 98
    6.8 Edge Defense 99
    6.9 Stages of Play 100
    Summary 102

7   Ladders 103
    7.1 Ladder Basics 103
    7.2 Ladder Formation 105
    7.3 Ladder Escapes 106
    7.4 Ladder Escape Forks 111
    7.5 Ladder Escape Foils 112
    7.6 Getting Off Ladders 119
    7.7 Partial Ladder Escapes 120
    Summary 125

8   Algorithmic Board Evaluation 127
    8.1 The Algorithm 127
    8.2 An Example 133
    8.3 The Need for Groups 141
    8.4 Optimizations 144
    8.5 Features of the Algorithm 145
    Summary 146

9   Opening Play 147
    9.1 Opening and Swapping 147
    9.2 Common Opening Strategies 156
    9.3 Adapt to the Situation 167
    9.4 Even-Sided Boards 169
    Summary 170

10 Strategy III: Advanced 173
    10.1 Multiple Threats Per Move 173
    10.2 Don't Provide Forcing Moves 180
    10.3 Ladder Handling 184
    10.4 Looking Ahead 196
    10.5 Overall Game Plan 202
    Summary 204

11 Annotated Sample Games 205
    11.1 Notation 205
    11.2 Ladder Escapes Denied 206
    11.3 Premature Resignation 211
    11.4 How Not to Play 217
    11.5 Another Point of View 219

12 Strategy IV: Essential 225
    12.1 Opening Play 225
    12.2 Start Blocking at a Distance 227
    12.3 Bridges 227
    12.4 Play Defensively 228
    12.5 Edge Templates 229
    12.6 Forcing Moves 230
    12.7 Ladders 231
    12.8 Spanning Paths 235
    12.9 Multiple Threats Per Move 237
    12.10 Looking Ahead 237

13 Hex Puzzles 239
    13.1 Previously Published 239
    13.2 Original 242

14 Conclusion 251

15 References 253
    15.1 Publications 253
    15.2 Online Resources 257

Appendices

A Solutions to Puzzles 261

B Some Notes on Berge's Hex Problem 281

C Sample Games 287

D Proofs 303
    D.1 One Player Must Win 303
    D.2 First Player to Win 305

   D.3 Acute Corner is a Losing Opening 305
   D.4 First Player Loses on n*(n + 1) Board Playing Wide 306
   D.5 No Simultaneously Opposed 0-Connected Spanning Paths 306

E Hex Variants 309
   E.1 Variants on the Hex Board 309
   E.2 Other Hexagonal Connectivity Games 313
   E.3 Non-Hexagonal Connectivity Games 319
   E.4 Tile-Based Connectivity Games 322
   E.5 Mobile Pieces 324
   E.6 Three-Dimensional Connectivity Games 325

F Blank Hex Boards 327

G Polyhexes 347

H Hex Programs 351

I Glossary 353
   I.1 Terms 353
   I.2 Symbols 356
   I.3 Path Algebra 357
   I.4 Move Notation 357

Index 359

Preface

Hex is an abstract board game that has fascinated mathematicians with its beauty and surprising difficulty since it was first invented over half a century ago. It's a seminal game that has inspired many variants over the years, some of which have achieved greater fame than Hex itself. Hex has recently enjoyed a surge of popularity and grown from its specialist niche to become recognized as one of the world's classic board games. However, there have been to date no books specifically about this intriguing game.

Hex Strategy is a comprehensive look at the game, from its historical origins and mathematical underpinnings to discussions of advanced playing techniques. It draws together into a single reference salient points from the many scattered sources of Hex literature currently available. Key concepts related to board analysis are defined, and it is hoped that this book will go some way to establishing a common vocabulary from which further research into Hex may develop.

This is first and foremost a book on strategy aimed at giving the reader sufficient knowledge to play Hex at any level desired. It's structured around four main chapters on strategy, organized sequentially from basic concepts through to more involved tactical considerations. The reader eager to improve their game as quickly as possible may wish to start at Chapter 14 for a brief summary of essential strategies, then refer back to relevant chapters for a deeper discussion of the points involved.

Due to the nature of the game, it is tempting to take a purely technical approach when discussing Hex. However, every effort has been made to make this book accessible to as wide an audience as possible while maintaining its technical accuracy. Abstract board game players, recreational mathematicians, AI programmers, or simply those who enjoy games and puzzles should all find something of interest within these pages.

The main goals in writing *Hex Strategy* were to introduce new players to Hex, improve existing players' enjoyment of the game, and to pin down the key strategies involved and describe them precisely in a single volume.

Structure of the Book

Hex Strategy is sequential in nature. Basic concepts are introduced and developed into increasingly involved discussions as the chapters progress. For instance, the chapter on opening plays requires some prior knowledge, and so occurs towards the end of the book rather than at the start where one might expect.

The following chapter summaries describe the logical flow of the book:

• *Chapter 1: Introduction* describes the game of Hex, including its rules, history, and a brief survey of existing literature.

• *Chapter 2: Adjacency and Connectivity* covers the basic concepts of adjacency and connectivity between pieces.

• *Chapter 3: Strategy I: Basic* outlines several strategies that can be employed immediately without a deep understanding of the game.

• *Chapter 4: Groups, Steps, and Paths* introduces concepts that allow the accurate analysis of players' overall connectivity across the board, including a simple path algebra.

• *Chapter 5: Templates* introduces predefined template patterns that describe the connectivity of players' pieces relative to each other and the closest home edge based on path analysis.

• *Chapter 6: Strategy II: Intermediate* incorporates material from preceding chapters to describe more involved points of strategy that require some understanding of the game.

• *Chapter 7: Ladders* introduces ladders and ladder escapes, probably the most important and involved Hex concepts.

• *Chapter 8: Algorithmic Board Evaluation* describes an algorithm that gives an accurate measure of each player's connectivity across the board using paths, templates and ladders.

• *Chapter 9: Opening Play* demonstrates key points of opening strategy and goes through some common opening sequences.

• *Chapter 10: Strategy III: Advanced* shows how the previously described points of strategy should be combined and employed in the context of the overall game to develop strong plays, through the use of example board situations.

• *Chapter 11: Annotated Sample Games* steps through a few noteworthy games move-by-move, pointing out relevant highlights.

• *Chapter 12: Strategy IV: Essential* encapsulates the most important Hex strategies in a nutshell. The reader impatient to explore the game's strategic aspects as quickly as possible may wish to start here.

• *Chapter 13: Hex Puzzles* provides a number of Hex puzzles both abstract and based on actual game play.

• *Chapter 14: Conclusion* briefly describes the overall work in context.

- *Chapter 15: References* lists works cited in this book and other relevant material.

- *Appendices* include solutions to Hex problems, a collection of sample games, blank Hex boards, and other supplemental material that may assist the reader's pursuit of Hex excellence.

Acknowledgments

Thanks to Richard Rognlie for maintaining the free play-by-email server Gamerz.NET which has proved an excellent forum for many games including Hex, and without which this book would not have been possible. Thanks also to those players who accepted my endless requests for games and discussions of strategy over the years, especially Leonid Gluhovsky, Patrick Mouchet, Thomas Hayes, Quentin Neill, Michelle Sporcic, Kevin Walker, David Boll, and Chris Lusby Taylor.

Leonid Gluhovsky and John Tromp deserve special mention for their meticulous proof-reading and general help with the book, and suggesting several corrections and improvements which were all spot on.

Thanks to Aviezri Fraenkel and David Singmaster for help tracking down Piet Hein's 1942 *Politiken* articles, and Mogens Esrom Larsen for going to the trouble of providing copies. Patrick Mouchet also provided several interesting articles related to Hex and other useful information. Thanks to Adrian Secchia for discussions on geometric aspects of the hexagonal board, and Frederic Maire for sharing his enthusiasm for the game.

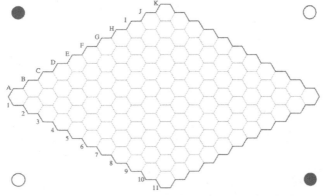

Introduction

This chapter provides a brief introduction to the game of Hex: its rules, history, and nature. A summary of interesting results from mathematical research inspired by Hex is presented. Various board styles on which it can be played are described.

1.1 The Game of Hex

Hex is an abstract board game in which two players compete to build a connected chain of pieces across opposite sides of the board. Its appeal to mathematicians lies in the puzzling nature of the game; its rules are among the simplest of any board game, yet to play it well is exceedingly difficult.

The Hex board is a hexagonal tiling of n rows and m columns arranged in a rhombus shape. Usually $n = m$ with 11x11 being the widely accepted standard board size. Examples in this book refer to an 11x11 board unless otherwise stated.

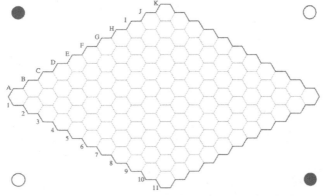

Figure 1.1. An 11x11 Hex board ready for play to begin.

Figure 1.1 illustrates a typical 11x11 Hex board ready for play. White's *goals* are the bottom left and top right edges marked with white pieces, and Black's goals are the top left and bottom right edges marked with black pieces. The four corner hexagons straddle edges of both colors and may be used by either player to connect.

The alphanumeric board notation used in this books describes cells on the Hex board by letter along White's direction of play (columns) and by number along Black's direction of play (rows). For instance the central hexagon is labeled F6. The term *point* refers to a board cell or hexagon, which can be either *empty* or *occupied* by a piece.

1.2 Rules of Play

The game is played by two opponents, Black and White. Each player owns the two opposite edges of the board that bear their color. The board is initially empty of pieces.

The rules of the game are simple:

•   *Players take turns placing a piece of their color on an unoccupied hexagon.*

•   *The game is won when one player establishes an unbroken chain of their pieces connecting their sides of the board.*

A game of Hex can never end in a tie. If one player completes a connection between the sides of their color, then the opponent is prevented from completing their connection. Figure 1.2 illustrates a completed game that has been won by White.

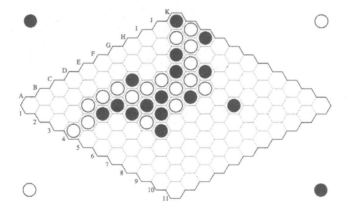

Figure 1.2. A game won by White, whose sides are connected by a chain of white pieces.

The player to move first has a huge (winning) advantage if they are allowed to make a strong opening play. An additional rule is often used to reduce this first move advantage:

•   *The player to move second has the choice of swapping colors, effectively stealing the first player's move.*

This is called the *swap option*, and it's recommended that all games be played with this additional rule. This is a good way of ensuring that the first player does not make too strong an opening move.

There is no universally accepted starting color; either Black or White may start the game. The player to move first may be chosen by the toss of a coin, by mutual agreement, or according to the outcome of previous games. It's common for opponents to engage in *dual challenges* (two games of Hex started in parallel, with each player making the first move in one game) to average out the first move advantage.

1.3 History

Hex was first invented by Danish mathematician Piet Hein. The first published description of the game can be found in an article by Hein in the December 26, 1942 edition of the Danish newspaper *Politiken*, in which he describes a game called Polygon. The 11x11 hexagonal board and the rules described by Hein are identical to those of Hex today.

The game was also invented independently by American mathematician John Nash in 1948 while he was a graduate student at the Princeton mathematics department [Nasar 1994]. This version of the game was named Nash in his honor, and sometimes called John due to the tendency of students to play it on hexagonal bathroom tiles. Thus Hex can boast an exceptional pedigree—its inventors are among the most acclaimed mathematicians of this century. In fact Nash recently became the 1994 Nobel Laureate for Economics for his pioneering analysis of equilibria in the theory of non-competitive games. An interesting account of Nash's life is given in [Nasar 1998].

The game has always been a favorite among students of mathematics, but was exposed to a wider audience with Martin Gardner's *Scientific American* article of the late 1950s. Parker Brothers marketed the game in the early 1950s under the name Hex [Nasar 1994]. This, in conjunction with Gardner's popular article, was probably the point at which the game was standardized to become what is known as Hex today.

In 1968 Piet Hein marketed the game under the name Con-Tac-Tix, and sold it as a 12x12 diamond shaped wooden board with a lattice of holes in which black and white pegs were placed [Finn 1998]. It has been produced by a number of different game companies over the years, and is often sold as a pencil-and-paper game played on a pad of 11x11 board diagrams. Such a game was published by Four Connections in 1974 under the name Polygon, using the square lattice design described in Section 1.4.3 [Finn 1998].

Hex has constantly drawn the attention of serious mathematical researchers and has yielded some interesting results (see Section 1.5) but has withstood attempts to solve it or to devise a winning strategy for boards greater than 7x7, beyond the fact that proving a winning strategy does exist for the first player. As far back as 1953 Claude Shannon implemented an analog Hex playing machine that played a reasonable game, and recognized its suitability for digital implementation [Shannon 1953]. Peter Arnold [1985] also observed that Hex was a prime candidate for computer implementation, but until recently the results have not been very impressive. This is possibly due to Hex's limited exposure but also hints at the extraordinary complexity of the game, which is perhaps on a par with computer Go in terms of difficulty of analysis.

There has been a recent resurgence in interest in Hex as a game played remotely across the Internet, where it is enjoyed by an increasing number of enthusiasts. See Section 15.2 *Online Resources* for a list of web sites that provide online games, introductory tutorials, Hex playing programs and other material related to the game. Two online forums for playing Hex deserve special mention: Gamerz.NET which provides a play-by-email service [Rognlie 1996], and PlaySite which allows interactive games between remote players [Go2Net 1999].

1.4 Nature of the Game

Hex belongs to the class of *two-person zero-sum finite deterministic games of strategy*. Melvin Dresher defines these terms in *Games of Strategy: Theory and Applications* [1961] as follows:

•   *game: A competitive environment described by a set of rules.*

•   *two-person: A game played by two opponents.*

•   *zero-sum: Players make payments only to each other (one player wins and the other player loses).*

•   *finite: The game has a finite number of moves and a finite number of choices available at each move.*

•   *deterministic or perfect information: A game in which players move alternately and at each move are completely informed about the previous moves. This definition can be expanded to include the fact that both players also have perfect knowledge about the current choice of moves and possible continuations of play.*

•   *strategy: A plan formulated for playing the game from beginning to end that results in a win. It has been proven for Hex that a winning strategy exists for the first playing, though the exact strategy is not known.*

Dresher discusses such games in terms of *payoff matrices*, and shows how *saddle points* within these matrices point to optimal strategies or winning plays. Unfortunately the payoff matrix for Hex becomes impracticably large except for trivial board sizes.

Sergio Antoy further narrows the classification of Hex into the subgroup of *positional board games*, that is, games in which players take turns placing pieces on a board [1987].

Hex can also be described as an *abstract game*, which Michael Keller defines as "a two-player game of pure strategy, with no chance elements, no hidden information, and alternate (not simultaneous) play" [1998]. Games with more than two players are not included in this category due to elements of diplomacy, coalitions, and partnership communication which may muddy the strategic waters.

David Parlett places Hex in context of the overall history of board games as belonging to the class of *Games of Linear Connection*, within the more general category of *Space Games* [1999]. He describes Hex as a classic of its type.

Some aspects of Hex warrant further discussion: *complexity, clarity* and *determinism*.

1.4.1 Complexity

A solution (winning strategy) for Hex has eluded researchers for over half a century. This is largely due to the combinatorial complexity of the game. Hex played on larger boards is comparable in complexity to Go, another game that is notoriously difficult to analyze.

In order to understand just how complex Hex is, let's define some basic concepts. *valid board position* is a combination of Black and White pieces upon the board that could have arisen during normal play. A valid board position is *unique* if it is not similar to any other rotation. Board positions are similar to their 180 degree rotated counterparts, but reflections do not apply on the Hex board. *Board situation* is another term for board position.

A board position is valid if:

• *the difference in the number of White and Black pieces on the board is zero or one (assuming that play began on an empty board), and*

• *it is a win for either player then play stopped immediately upon the win.*

The empty board is considered a valid board position.

Figure 1.3 shows all seventeen unique valid board positions of a 2x2 board by way of example. The top left board is empty and shows the goals of each player. It is convenient to assume that Black moves first—there is no loss of generality.

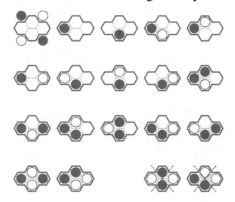

Figure 1.3. The seventeen unique valid and two invalid board positions for the 2x2 board.

The two boards at the lower right of Figure 1.3 have been struck out, as they could not have arisen during normal play and are therefore invalid. Black would have won the game with the third move on each of these boards, making White's fourth move redundant.

John Tromp points out that the total number of unique board positions U for any nxn board is given by:
$$U = \frac{X + S}{2}$$

At the ith move there are $\lceil i/2 \rceil$ of the first player's pieces and $\lfloor i/2 \rfloor$ of the second player's pieces on the board, where $\lceil i/2 \rceil$ denotes $ceil(i/2)$ and $\lfloor i/2 \rfloor$ denotes $floor(i/2)$. These terms are explained indetail in Weisstein (1999).

The values X and S can then be calculated as follows:

$$X = \sum_{i=0}^{n^2} \left[\binom{n^2}{\lceil i/2 \rceil} \times \binom{n^2 - \lceil i/2 \rceil}{\lfloor i/2 \rfloor} \right]$$

where $\binom{n}{k}$ denotes the binomial *coefficient* defined by $n! / (k! (n-k)!)$. If n is even then:

$$S = \sum_{i=0}^{n^2/4} \left[\binom{n^2/2}{i} \times \binom{n^2/2 - i}{i} \right]$$

else for odd values of n:

$$S = 2 \times \sum_{i=0}^{\lfloor n^2/4 \rfloor} \left[\binom{\lfloor n^2/2 \rfloor}{i} \times \binom{\lfloor n^2/2 \rfloor - i}{i} \right] + \sum_{i=0}^{\lfloor n^2/4 \rfloor - 1} \left[\binom{\lfloor n^2/2 \rfloor}{i} \times \binom{\lfloor n^2/2 \rfloor - i}{i+1} \right]$$

The result of this calculation is the total number of unique board positions for an $n \times n$ board, which may contain *invalid* positions in which play has continued after a win, as shown in figure 1.3. However, it is a useful indicator of the order of complexity that can be expected for a given board size.

An estimate of the upper limit on unique valid board positions is given by $\lceil 3^{n^2}/2 \rceil$ as each point may exist in either of three states: Empty, Black or White. However this exponential approach gives a gross overestimate and should only be used in cases where the more exact formula is not suitable.

The exact number of unique valid board positions were determined for $n = 1$ to 4, as shown in Table 1.1:

Board Size n	Valid Positions	Upper Bound	Exponential Estimate
1x1	2	2	2
2x2	17	19	41
3x3	2,844	3,050	9,842
4x4	4,835,833	5,083,443	21,523,361

Table 1.1. Exact number, upper bound, and exponential estimate of valid board positions for n =1 to 4.

The upper bound on valid board positions for the standard 11x11 board is roughly an impressive 2.38×10^{56}, and for the 14x14 board increases to approximately 1.14×10^{92}. This degree of combinatorial complexity has serious implications for players and Hex-playing computer programs. The complete game tree is too large to be processed by standard tree-searching techniques even for mid-sized boards, so additional methods must be found to prune the decision tree. As with Chess and most other abstract board games, the expert player will perform a very deep search on a few select moves, and ignore mediocre ones. The following chapters of this book introduce concepts about the game that will (hope-fully) allow the reader to make this distinction, and collapse entire subsections of the game tree to a single accurate evaluation, making the search easier.

Nine Men's Morris, another abstract board game in the same class as Hex, was re-cently solved and shown to be a draw [Gasser 1996]. Gasser's approach was to compute a database of 10^{10} endgame positions, then use an 18-ply alpha-beta search from the current board position to reach these known states. This approach was feasible as the Nine Men's Morris board is relatively small at only 24 board points. A similar approach to Hex would not work due to its larger playing surface and much greater *branching factor* (number of move choices per turn). A related topic is discussed in Appendix G, Polyhexes.

1.4.2 Clarity

If a game's worth can be estimated by its strategic depth versus rule complexity, then Hex provides excellent value. It's extraordinarily complex yet with a rule set among the simplest of any game possible.

However, estimating the strategic depth of a game is not as simple as describing the size of the complete game tree. As Robert Abbott points out in relation to his game Ultima, the depth of a game depends not so much on the *size* of the game tree as on *how far a player can see* down the game tree [1988].

This concept of *clarity* essentially describes the amount of certainty with which a player can plan ahead and formulate strategies. One problem that Abbott observed in Ultima is lack of clarity due to the fact that different pieces employ different methods of movement and capture; the potential threats posed by some pieces are clear and by some pieces are not.

Hex, on the other hand, has very good clarity. Each piece is of uniform strength, and it is often possible to plan 15-20 moves ahead with reasonable certainty. The *cut/short* nature of Hex is very well defined; each player either completes their connection (*short*) or is prevented from doing so (*cut*). This analogy of the Hex board as an electric circuit is explored further in relation to the game of Gale in Section 1.5.1.1. Some games derived from Hex but with more complex rule sets such as Havannah (see Appendix E.2) suffer somewhat from lack of clarity.

1.4.3 Determinism

Hex is a *deterministic* game in which both players have perfect knowledge of:

* *previous moves,*

* *the current choice of moves, and*

* *possible continuations of play.*

Chance plays no part in Hex. Both players have perfect knowledge about the game, and once a piece is played it remains in that position for the rest of the game. There is a fine distinction between chance and luck, however. Luck may be involved if a player makes a random move beyond their understanding of the game that later proves useful, but good players soon develop an instinct for the potential consequences of any move.

For this reason Hex is not amenable to another method of machine learning that has recently proved successful for other board games, Temporal Difference (TD) learning. TD learning involves neural networks that learn through the results of games played against themselves. It has been successfully applied to Backgammon [Tesauro 1995], Chess [Thrun 1995] and Othello [Walker et al. 1993], among other applications.

However, network learning requires some degree of randomness in the games being studied, so that it can learn from both good and bad plays to investigate the game space thoroughly. Gerald Tesauro explains the importance of this in [Peterson 1997]. This variation is missing from Hex and attempts to introduce randomness result in board positions that do not usually arise in normal play. Although Chess and Othello are also deterministic they often undergo dramatic changes in board state during a game, whereas pieces on the Hex board remain static.

A different approach to providing the learning system with sufficiently varied data is to feed it large numbers of actual games from expert game databases. This approach has been applied to Go [Stoutamire 1991] with a database of 55,000 moves but is not without its drawbacks; the system may pick up players' stylistic idiosyncrasies. For instance a system trained on expert play may make poor decisions against a beginner who makes unorthodox moves. In any case, the largest known Hex database contains less than 1,000 games, many of them non-expert—not nearly sufficient for training purposes.

It appears that purely connectionist methods are insufficient for implementing an effective Hex player, and that some knowledge-based analysis combined with selective search is required. Queenbee, a Hex-playing program currently being developed by Jack Van Rijswijck, employs some machine learning techniques to play a strong game [Van Rijswijck 1999].

1.5 The Shannon Switching Game

Hex is a special case of a more general type of game called the *Shannon Switching Game*, invented by Claude Shannon in the 1950s. This is played on an undirected graph with two distinct vertices s and s' (*source* and *sink*) as shown in Figure 1.4.

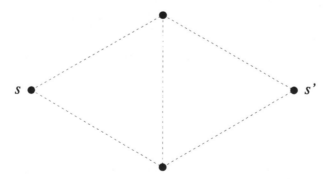

Figure 1.4. A simple graph with two distinct vertices s and s'.

The game is played between two opponents:

- *Short: attempts to complete a connection (short circuit) between s and s', and*

- *Cut: attempts to prevent this by cutting Short's connection.*

Each edge in the graph lies between two vertices and exists in one of three states:

undecided, *cut*, or *shorted*, as shown in Figure 1.5. An edge that is cut completely disconnects its vertices, and an edge that is shorted connects its vertices solidly.

The graph is analogous to an electric circuit, and the game's terminology and goals belie its origins in electrical engineering research. It can be played in two distinct forms: on the *edges* and on the *vertices* of the graph.

Figure 1.5. Edge states: undecided, cut, and shorted.

1.5.1 Played On the Edges

The rules for playing on the edges are shown in Figure 1.6. Cut removes an undecided edge per turn, and Short solidifies or short-circuits an undecided edge per turn. Play alternates between Cut and Short.

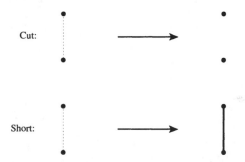

Figure 1.6. Rules for playing on the edges.

Figure 1.7 shows two complete games played on the simple graph shown in Figure 1.4. Short plays first and wins the game in the top row, while Cut starts and wins the game along the bottom row.

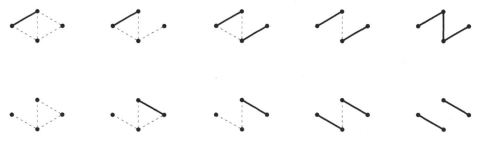

Figure 1.7. Short plays first and wins (top row) and Cut plays first and wins (bottom row).

Notice that the same sequence of moves has been made by each player and gives them the win with first move. The game's strategy is equivalent for both players.

1.5.1.1 The Game of Gale

A particular version of the Shannon Switching Game played on the edges is shown in Figure 1.8. This game is commonly called Gale in honor of its inventor, David Gale.

The game starts with all vertices on the top row solidly connected to vertex s and all vertices on the bottom row solidly connected to s'. This is a way of specifying that short need only connect any vertex in the top row with any vertex in the bottom row to win the game. The Figure on the right shows a game won by Short who has established a connection between a vertex on the top row with a vertex on the bottom row.

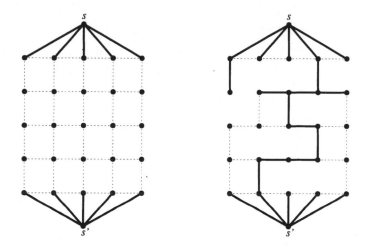

Figure 1.8. The Gale board and a completed game.

Gale was commercially released under the name Bridg-It with the added limitation that only 10 cuts and 10 shorts may be made per game—if the game is not won after their placement each player must move an existing cut/short respectively per turn.

Analyses by Lehmann in 1964 led to very specific strategies for Gale. A simple and elegant solution to the game based on a pairing strategy was developed by Oliver Gross, and is described by Martin Gardner [1966].

1.5.2 Played On the Vertices

The Shannon Switching Game becomes much harder when played on the vertices as demonstrated by Even & Tarjan [1976]. This is the category to which Hex belongs.

The same format for describing the game on the edges can be used for vertices, but with the modification shown in Figure 1.9. Now players choose one vertex per move rather than one edge. Each edge is treated as being composed of two *half-edges*. All half-edges

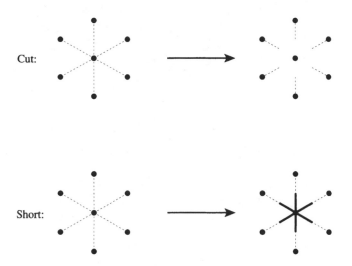

Figure 1.9. Rules for playing on the vertices.

containing Cut vertices are removed, while all half-edges containing Short vertices are solidified. This allows us to describe the Hex board as a Shannon network. Figure 1.10 shows a 5x5 Hex board described in this fashion.

Notice that the graph is composed of *major* vertices (large dots) and *minor* vertices (small dots). Players may choose one of the major vertices each turn. The minor vertices are intermediate points used to visually denote the division of half-edges, and are not valid moves in the game.

Again, the solid connections leading from each vertex along the top left edge to vertex s and leading from each bottom right vertex to s' indicates that Short only has to form a

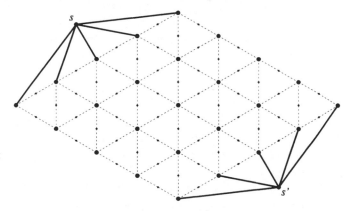

Figure 1.10. Shannon network corresponding to the 5x5 Hex board.

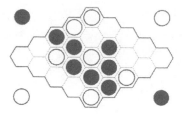

Figure 1.11. Example 5x5 game of Hex.

connection between a single vertex on each of these edges to win. Conversely, Cut only has to establish a break between a single vertex on each of their edges to win. By way of example a completed 5x5 game of Hex is shown in Figure 1.11.

The same game on the Shannon network is shown in Figure 1.12.

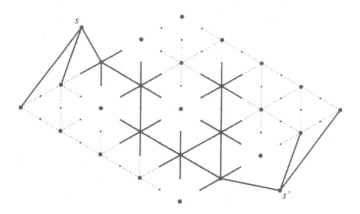

Figure 1.12. The same game shown on the Shannon network.

In 1951 Claude Shannon built an electrical Gale-playing device which he called the Birdcage [Gardner 1961]. The analog device consists of a network of resistors that corresponds to edges in the graph. On their move each player physically applies the concept that their name specifies: Cut opens a connection by removing the corresponding resistor from the circuit, and Short closes a connection by short circuiting the corresponding resistor.

The game is won by Cut when the current across the circuit is cut off entirely, and won by Short when the resistance across the circuit drops to zero. During play the resistor across which the maximum voltage occurred was chosen as the best move. This strategy proved sound and the device played a very good game, winning almost all of its matches if allowed to play the first move.

Shortly afterwards Shannon built a Hex-playing device based on similar principles [Shannon 1953]. This was composed of a resistor network corresponding to a Hex board, with one player's two sides made positive and the other player's two sides made negative.

Each player's piece was given a charge of opposite polarity to their sides. Resistance saddle points indicated the best move at each turn when current was applied across the board.

The Hex device played a reasonable game and defeated most opponents if given first move. Shannon observed that surprisingly the device was good at general positional play but weaker in its end game combinatorial play.

One problem with the analog approach is that it is difficult to encode the network strategies and formations of play that an experienced player will use as a matter of course. It's doubtful that the analog Hex-playing device described by Shannon would fare very well against an expert player. However, the idea of modelling the Hex board as an electric circuit did inspire Vadim Anshelevich to develop the algebra of virtual connections that forms the basis of his excellent Hex-playing program Hexy [Anshelevich 2000].

1.6 Current Literature

Although there have previously been no books devoted to Hex, a small body of literature has emerged from the pages of mathematics texts and journals, both recreational and serious. The appeal of Hex to all types of mathematicians will soon become apparent. This section provides a survey of current Hex literature in approximately chronological order, and describes the key points of each contribution. The survey is broad, and for more in-depth discussion of each topic the reader should consult the reference cited (Chapter 15).

Some mathematical theorems and concepts related to Hex are introduced briefly, but it is not necessary to understand these in detail in order to play Hex well. There are, however, two important results that the player should always keep in mind:

- *The game will always result in a win for one player (no draws).*

- *The first player has a winning line if they play without fault.*

These points will be discussed in detail in following chapters.

Piet Hein's 1942 article in the Danish newspaper *Politiken* describes the rules of Hex, which he originally called Polygon. The game's origins are themselves very mathematical; in nature: it occurred to him while contemplating the famous four-color theorem of topology, and was introduced to students in a lecture at the Niels Bohr Institute for Theoretical Physics in Copenhagen [Gardner 1959]. Hein's original article was followed by a series of pieces on Polygon including competitions to solve Polygon problems. One of these problems is given in Section 13.1. These articles are in Danish.

Claude Shannon [1953] offers a brief but tantalizing description of a Hex-playing machine that he had constructed in his article "Computers and Automata." The machine chose moves based on saddle points in the potential field across a resistance network that reflected the current board state. It played a reasonably strong game, good at general positional play but weaker at end game combinatorial play—not what one would expect from an analog implementation.

Martin Gardner's Scientific American column "The Game of Hex", reprinted in his 1959 book *Mathematical Puzzles and Diversions*, was the breakthrough article that brought Hex from the mathematical domain and introduced it to the world. It describes the history of the game, introduces some interesting mathematical properties, and presents some Hex problems. Nash's famous *strategy stealing* proof that the first player has a winning line is described. This proof is quite intriguing in that it shows that the first player has a winning play, but does not indicate what that play might be.

John Pierce [1961] describes the meaning of the terms *theorem* and *proof* using Hex as an example in his book *Symbols, Signals and Noise*. The theorems he examines are:

- *either one player or the other must win, and*

- *the player with the first move should win with optimal play.*

Pierce demonstrates that any board entirely filled with pieces must contain a winning chain for one of the players, and offers a version of Nash's strategy stealing proof.

Another proof for the theorem that the opening player has a winning line is given by Anatole Beck [1969] in his book *Excursions Into Mathematics*. The rather detailed proof makes use of a version of Hex called "Beck's Hex" which is identical to standard Hex, except that the second player dictates where the first player must make their opening move. Beck also shows that opening in either acute corner hexagon (A1 or K11) is a losing play. This is discussed further in Section 9.1.3.1. The swap option essentially makes Hex a friendlier version of Beck's Hex.

Beck notes that already a number of mathematicians had attempted and failed to discover an explicit winning strategy for Hex, and "whoever solves it will achieve thereby a worldwide, if fleeting, renown." Hex remains unsolved to this day.

A review of Hex from the game-player's perspective is presented by Rodney Headington in his 1973 article "The Game of Hex" in *Games & Puzzles* magazine.

Ronald Evans published two short articles on Hex in the *Journal of Recreational Mathematics:* "A Winning Opening in Reverse Hex" [1974], and "Some Variants of Hex" [1975-76]. The first article introduces the game of Reverse Hex, in which the object of both players is to lose. Evans builds on Beck's analysis to show that the first player can force a loss, and hence win Reverse Hex, by playing their opening move in an acute corner. The second article introduces a number of variations on standard Hex that are described in detail in Appendix E.

By this time Hex was becoming recognized as a "classic" board game, and was included in an increasing number of board game compendiums. Robert McConville makes a brief note of Hex in *The History of Board Games* [1974] including a description of the winning moves on 4x4 and 5x5 boards, but fails to elaborate for larger boards. A more in-depth discussion of Hex complete with an annotated sample game appears in *The Illustrated Book of Table Games* [Arnold 1975]. Arnold points out that the lack of Hex literature is to the advantage of the player who prefers over-the-board ability rather than prior study.

Other compendiums such as *Waddington's Illustrated Encyclopedia of Games* [1975] and *The Book of Games* [Arnold 1985] include brief single-page descriptions of Hex with little discussion of strategy. *The Book of Games* contains a stripped-down version of Arnold's previous discussion on Hex, who observes that "no doubt it will attract the attention of programmers ere long."

Schensted and Titus describe the game of Y in detail in *Mudcrack Y & Poly-Y* [1975].

Y is a hexagonal connectivity very closely related to Hex, and Schensted and Titus actually show that Hex is a special case of Y. Although the book consists primarily of hand-drawn boards for the reader's use, it does contain an excellent section on Y strategy that is largely relevant to Hex. The relationship between Y and Hex is discussed in more detail in Section 2.3.2.

Even and Tarjan present a detailed analysis of Hex in their paper "A Combinatorial Problem Which Is Complete in Polynomial Space" [1976]. They propose a generalization of Hex, which, as the name of the paper suggests, was shown to be complete in polynomial space, or PSPACE-complete. This essentially means that it is difficult to solve Hex without constructing the entire game tree, which is for all practical purposes impossible given its combinatorial complexity. This result puts Hex in the "extremely hard" basket, and indicates why a solution to the game continues to elude researchers. They note that games are harder than puzzles as the initiative shifts between players with each turn.

Concerned at the length of existing proofs of the first player win, David Berman proposes a simple inductive proof that the game cannot be drawn in his article "Hex Must Have a Winner: An Inductive Proof" [1976]. Two letters to the editor by Paul Johnson and Daniel Zwillinger in the same issue refer to Berman's article and outline even simpler proofs.

Morton Davis's paper "On Artificial Machine Learning: Some Ideas in Search of a Theory" [1976] describes an algorithm for machine learning based on manipulation of evaluation functions, and applies it to the problem of learning to play Hex and other various finite, two-person, zero-sum, perfect information games. In relation to Hex, this algorithm improved upon random play but did not reach the standard of a human beginner. This work is also discussed in his 1986 paper "Computer Learning of Parlor Games."

David Gale's celebrated 1979 paper "The Game of Hex and the Brouwer Fixed-Point Theorem" offers yet another elegant proof that a completely filled board must be a win for either player, and discusses the equivalence of this result and the famous Brouwer Fixed-Point Theorem. One of the most often-cited Hex references, it has led to further avenues of research. This paper also examines n-dimensional Hex.

Two short articles related to Hex problems were published in 1981:

•   *Duane Broline's "Kriegspiel Hex" [1981] introduces a variant of Hex in which both players can only see their own moves. This paper also provides some analysis on a 3x3 board.*

•   *Claude Berge's "Some Remarks about a Hex Problem" [1981] proposes a deceptive Hex puzzle (see Appendix B).*

Stefan Reisch uses a different approach to Even and Tarjan to show that Hex is PSPACE-complete in his "Hex ist PSPACE-vollstandig" [1981]. This paper is in German (as you may have guessed).

Hex receives a brief mention in Berlekamp, Conway, and Guy's excellent 1982 book on strategic gameplay, *Winning Ways for Your Mathematical Plays*. No Hex strategy as such is discussed, but some mathematical aspects of the game are mentioned. A strategy stealing proof of first player win is applied to the related game Bridg-It and its generalization the Shannon Switching Game.

Sergio Antoy's paper "Modeling and Isomorphisms of Positional Board Games" [1987] examines Hex along with other positional two-person board games. He identifies conditions necessary for establishing the isomorphism of two boards, and provides a model for the description and analysis of these games that is "more abstract than its physical characteristics such as size and dimensionality" [Antoy 1987].

Martin Gardner recaps some of the Hex literature in his 1998 books *Time Travel and Other Mathematical Bewilderments and Hexaflexagons and Other Mathematical Diversions*. The latter title is actually a reprint of his 1959 book that contained the groundbreaking article "The Game of Hex" with some new material included in the section "Afterword, 1988".

John Beasley describes Nash's strategy stealing proof in a section aptly titled "When you know who, but not how" in his book *The Mathematics of Games* [1989]. Beasley explains how the same argument can be applied to Bridg-It, though a straightforward winning strategy is known for first player [Lehman 1964].

Richard Guy briefly mentions that Hex is PSPACE-complete in conjunction with other conditionally intractable games in *Combinatorial Games* [1991].

Alpern and Beck move from rectangular Hex to cylindrical Hex and Hex played on twist maps on the annulus in their article "Hex Games and Twist Maps on the Annulus" [1991]. Gale's work relating Hex to Brouwer's Fixed-Point Theorem is discussed in this context.

Ken Binmore's *Fun and Games: A Text on Game Theory* [1992] contains several sections pertaining to Hex. Binmore explains some of the previous mathematical work on Hex (including Brouwer's Fixed-Point theorem and the proofs for no games tied and first player win) in a clear and most readable form.

Theonis Pappas includes a one-page description of Hex in his 1994 collage of mathematical curiosities The Magic of Mathematics. This book also contains a section on the related topics of polyhexes and polyhex puzzles.

Games of No Chance [Nowakowski 1996] does not contain any material specifically on Hex, but contains a remarkably complete bibliography on combinatorial games that incidentally includes several Hex references. Aviezri Fraenkel, in his introduction to the bibliography, makes the interesting observation that "the opponent can be very mean"; combinatorial games are harder than passive problems of similar complexity in that the opponent actively works to frustrate their solution.

Arratia-Quesada and Stewart further investigated PSPACE-completeness of Hex in their article "Generalized Hex and logical characterizations of polynomial space" [1997].

David Book's 1998 article "What the Hex" published in The Washington Post gives a good overview of the game and suggests some interesting variations. These include Hex played on semi-regular and irregular polygonal tilings, and even aperiodic tilings such as maps. Non-Regular Hex is discussed further in Section 2.3 and Appendix E, Hex Variants.

Hex is described in the section "Games of Linear Connection" in *The Oxford History of Board Games* by David Parlett [1999]. Parlett recognizes the importance of Hex in the context of board game development and lists several excellent Hex-inspired games.

Hex has recently found its way into the pages of the 1999 *CRC Concise Encyclopedia of Mathematics*, though the brief analysis of the game presented there is incorrect.

Some additional material on Hex, such as online documentation posted on the Internet, remains unpublished but worth mentioning. These include the excellent "Hex: Answers to Common Questions" [Boll 1994], which describes the game in detail, gives sound strategic advice, and provides an annotated sample game. This work contains the most complete discussion of Hex strategy to date, and contains enough information to enable the reader to begin playing at an intermediate level. Bert Enderton's "Answers to infrequently asked questions about the game of Hex" [1995] contains interesting facts related to Hex, including the fact that Hex has been solved up to board size 7x7, and some puzzles on smaller boards.

Jack Van Rijswijck provides excellent illustrated explanations of a couple of the Hex proofs mentioned above in his web site "Hex" [1998].

1.7 Board Representations

Hex may be played on a variety of board types. The boards described in this section are simply different representations of the same game, and do not affect the actual play.

Some players consider the rhombus board to be an inferior design due to the presence of two distinct types of corners (acute and obtuse), and because edge points near the acute corners are further from the center than other edge points around the board. However, this is little more than a question of aesthetic preference. The presence of distinct corner types allows a broader range of strategy, and the game is made richer by the intrinsic topology of the board rather than by manipulating the rules.

1.7.1 Variable Board Size

Although 11x11 is widely accepted as the standard board size, Hex may be played on any nxn board (or mxn board as discussed in Appendix E.1). The optimal board size for a given player largely depends on their standard of play; novice players may prefer to learn the basics on a smaller board, while more experienced players generally prefer larger boards such as 14x14 to 17x17. Boards ranging in size from 3x3 to 26x26 and suitable for copying are provided in Appendix F for the reader's benefit.

The smallest board is the trivial 1x1 board as illustrated in the leftmost Figure 1.13. This board has the serious limitation that only one move is possible; the opponent does not get to play a piece before the game is over.

Figure 1.13. Trivial board sizes 1x1, 2x2, and 3x3.

The 2x2 board is the smallest board size on which the opponent gets a guaranteed reply. However, the game will end on the next move unless the opening player plays diabolically badly, and the board does not provide any *interior* (non-edge) points and so is also of limited interest. The smallest board of interest is the 3x3 board shown in rightmost Figure 1.13. It allows a total of 2,843 unique valid board positions as described in Section 1.2.1 and contains an interior point p.

Hex has been solved for all boards up to and including 7x7 [Enderton 1995]; hence, experienced players will not find boards smaller than 8x8 very interesting for general play. Smaller boards are used primarily for tutorial purposes, establishing strategies that are projected to larger boards, or setting up Hex puzzles.

The 11x11 board is widely accepted as the *standard board*. All strategies discussed in this book are described with the 11x11 board in mind but can generally be extrapolated to larger board sizes. A subset of these strategies will apply to smaller boards (the smaller the board, the smaller the subset) but boards less than 11x11 generally do not provide a satisfactory game.

14x14 is a common choice for experienced players, and is hence described as the *expert board*. This size reduces the opening player's first move advantage and allows for a more complicated game, but does not blow out to marathon games as are common on the 17x17 or 19x19 sizes.

In general, the larger the board, the more room there is to play around a threat; hence each move is less decisive and the game more subtle. However, this is taken to extremes in the case of the 26x26 board (shown in Figure 1.14). 26 is a convenient upper limit to board size, as this is the maximum range of board labels using the alphanumeric notation. The 26x26 board is getting impractically large with an estimated upper bound of 1.7×10^{322} valid board positions.

Figure 1.14. 26x26: the largest recommended board size.

The 26x26 board is strictly for the ambitious player with a lot of time on their hands. Typically a game is resolved before a third of the board is covered, but occasionally a game may cover half the board or more. For an 11x11 board the average game will last about 40 moves. For a 26x26 board we can predict that an average game would require more than 200 moves.

1.7.2 Explicit Goal Labels

Figure 1.15 shows the original board design presented by Piet Hein in his 1942 article.

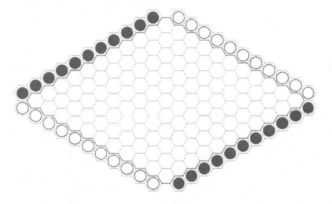

Figure 1.15. Empty board with explicitly labeled goals.

This board description is identical to the standard board shown previously, except that complete rows of each players' pieces are shown along each edge, removing any doubt as to which edge belongs to which player. Explicit goal labeling is useful for board analysis as connections to the edge are visible rather than implied, but has the disadvantage that the board appears cluttered before the game has even begun. The presence of extra pieces can be distracting to the player as the game progresses and the board fills even further.

For clarity, the more concise method of using single pieces of the appropriate color to indicate goal edges is used throughout this book.

1.7.3 Diamond Lattice

The board is often represented as a diamond shaped grid composed of equilateral triangles (Figure 1.16). Play occurs on the intersections of lines as in Go. There is otherwise no difference between play on this board and play on the standard board composed of hexagonal tiles.

The diamond lattice is easier than the hexagonal grid and is often used if the game is to be played with pen and paper. It also has the advantage of emphasizing point-to-point connections through lines that connect adjacent points. However, if a pad of hexagonal graph paper is available, this provides a suitable grid on which the standard hexagonal board may be drawn.

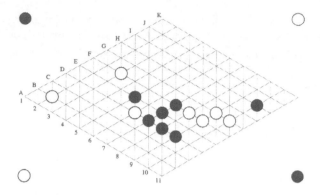

Figure 1.16. 11x11 diamond lattice board with a game in progress.

1.7.4 Square Lattice

The square lattice (Figure 1.17) is similar to the diamond lattice except that the points have been shifted so that rows and columns are parallel to the horizontal and vertical axes respectively. Again, play occurs on intersections. This board format is easier to draw than either the hexagonal tiling or diamond lattice but the game loses much of its attractive hexagonal quality. The connections between points are equivalent to connections on the hexagonal board. The square lattice is often used when the game is played in pencil-and-paper format, such as Polygon, marketed in 1974 [Finn 1998]. This is also the format proposed by John Nash and used by David Gale [Gale 1979].

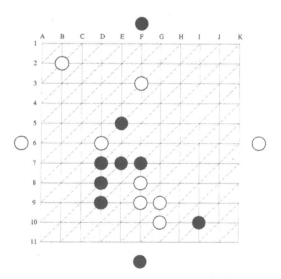

Figure 1.17. 11x11 square lattice board showing the same game as Figure 1.5.

The square lattice representation will be most familiar to Go players, and may ease their transition to Hex. However for Hex players familiar with a hexagonal layout the square lattice is generally confusing and less pleasant to play on.

1.7.5 ASCII Text Boards

The ASCII text format is used in situations where plain text characters are preferred over graphical representation. This includes most cases in which a computer is involved in playing the games such as via email correspondence, or if game records are to be interpreted and processed for computer analysis. The ASCII version is not as pleasant to use as graphical board representations and can become quite confusing, especially if board formatting is corrupted by the use of non-monospaced typefaces. However, the text format allows the game to be played on any text editor.

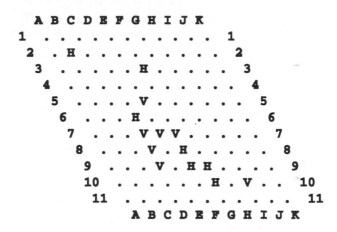

Figure 1.18. 11x11 ASCII text board suitable for email play and machine processing.

ASCII text format requires the board to be rotated slightly so that play occurs in the horizontal and vertical directions. This has led to the notation of vertical and horizontal players called V and H that replace Black and White as used throughout this book. Figure 1.18 illustrates the same game shown in 1.15 and 1.17 on an ASCII text board using V and H notation to denote players' pieces.

This is the format used for games played on the Gamerz.NET play-by-email server [Rognlie 1996], where it is the convention for V (Black) to start. This is the convention used in most examples throughout this book.

1.7.6 Scrambled Hex

Scrambled Hex is an interesting board format that does not alter the structure or topology of the underlying board, but rearranges the way it is presented to the player. Some players will enjoy solving the maze-like connections, while others may be irritated by the confusing layout. In any case it makes a difficult game that much more difficult.

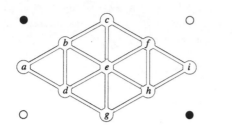 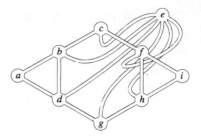

Figure 1.19. Graph representation of a 3x3 board and a knot embedding with one vertex displaced.

Figure 1.19(i) illustrates how a 3x3 Hex board can be represented as a *graph* with *nodes* corresponding to board points, and *edges* between adjacent points.

Figure 1.19(ii) demonstrates how a *knot* can be obtained from the graph by treating edges as solid three-dimensional objects that can be deformed in space. Both Figures are *isomorphic embeddings* of the same graph; that is, the underlying connections are equivalent in both cases [Adams 1994]. The second Figure is also an *isotopic knot* derived from the first Figure by applying a single *Reidemeister move*.

Figure 1.20 shows a more convoluted isomorphic embedding of the graphs in Figure 1.19, but is not necessarily an isotopic knot of either. That is, there is not necessarily any way to deform the knots shown in Figure 1.19 through space to derive the knot shown in Figure 1.20 without allowing edges to pass through themselves or each other (there may exist no sequence of Reidemeister moves).

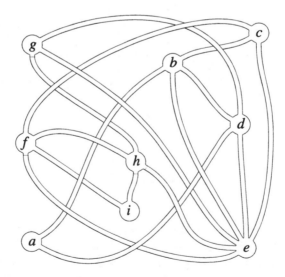

Figure 1.20. Convoluted embedding of the same 3x3 board.

However, isotopy is irrelevant to the knot's function as a Hex board provided that the underlying graph is an isomorphic embedding of a valid Hex board. The knot visualization with over-under weave pattern is a purely decorative effect to achieve a convoluted maze effect.

The greatest difficulty with Scrambled Hex lies in indicating which points are edge points without unduly cluttering the design. An additional connection leading from each edge point to its closest goal marker can solve this problem.

Some examples of interesting mazes derived from similar principles can be found in *100 Perceptual Puzzles* [Berloquin 1976].

Summary

The rules of Hex are simple and easy to learn. An optional rule that allows the second player to swap the opening piece is recommended to reduce the first move advantage. Hex is complex, purely deterministic, and has extremely good clarity of play.

Hex has attracted the attention of mathematicians for over 50 years due to its puzzling nature: its rules are among the simplest of any board game, yet to play it well is exceedingly difficult, and a winning solution proves elusive.

The main body of literature is composed primarily of papers from mathematics books and journals, and no previous book on Hex strategy has been published. Hex is widely recognized as a classic game, but is certainly not a household name.

Hex can be played without modification on a variety of board representations, provided that the underlying topology of the board agrees with that of a standard Hex board.

Adjacency and Connectivity

Understanding the connectivity between pieces on the hexagonal grid is central to understanding the game of Hex. After a brief discussion of the hexagonal coordinate system, this chapter defines some concepts related to piece adjacency, connectivity, and distance on the hexagonal grid. These provide the foundation for more detailed analyses of board situations and Hex strategy.

The suitability of the hexagonal grid for connectivity games of the Hex class is demonstrated by examining the strengths and weaknesses of related games played on various types of grid.

2.1 Coordinate Systems and Adjacency

Two cells on the Hex board are *adjacent* if they share an edge. Cells adjacent to point p in Figure 2.1 are shaded gray and are described as *p's immediate neighbors*.

Two pieces are adjacent or *touching* if they occupy adjacent cells, such as pieces a and b in Figure 2.1. Pieces c and d are non-touching and are separated by one step at point q. If White were to move at q then Black would have difficulty connecting c and d.

From Figure 2.2 it can be seen that cells on the rectangular grid have four immediate neighbors and cells on the hexagonal grid have six immediate neighbors. The use of the

Figure 2.1. Adjacent cells, touching pieces, and non-touching pieces.

traditional rectangular [*row, column*] system for labeling points on the Hex board can be seen as applying hexagonal adjacency generators within the rectangular grid as shown on the right of Figure 2.2.

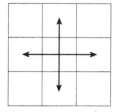 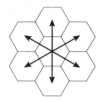 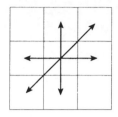

Figure 2.2. Adjacency generators: rectangular, hexagonal, and hexagonal generators in a rectangular grid.

A more intuitive way of visualizing this relationship devised by Alpern and Beck is shown in Figure 2.3 [1991]. The Figure on the left shows a square 11x11 grid with alphanumeric position labels. Each cell is adjacent to the four neighboring cells with which it shares an edge (rectangular generators).

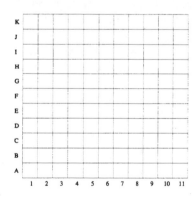 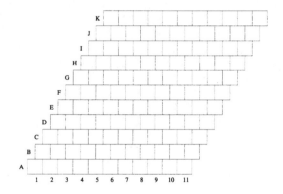

Figure 2.3. Square and skewed rectangular grids.

The Figure on the right is the same grid skewed such that each row is displaced by half a unit to the right relative to the row beneath it. Each cell is still adjacent to other neighboring cells with which it shares an edge, which for the skewed case is equivalent to the set of hexagonal generators. The skewed 11x11 grid is equivalent to the standard 11x11 Hex board.

A more in-depth discussion of hexagonal array grammars and adjacency generators can be found in [Aizawa 1989]. Paul Bourke [1996] describes an alternative hexagonal labeling system based on spiral coordinates.

2.1.1 Distance on the Grid

The *city block distance* between two points $[p_i, p_j]$ and $[q_i, q_j]$ on the rectangular grid is the minimum number of adjacent neighbors that must be traversed in order to reach one point from the other. This measurement is given by $dist_{rect} = |d_i| + |d_j|$ where $d_i = q_i - p_i$ and $d_j = q_j - p_j$. City block distance on the hexagonal grid (*beehive distance?*) can be computed from rectangular point labels by $dist_{hex} = max(|d_i|, |d_j|, |d_i + d_j|)$.

Figure 2.4 shows grid distances based on adjacent moves from a central piece for the rectangular and hexagonal grids. Distances shown indicate the number of adjacent moves required to reach each point from the central piece. The connectivity between two points on the hexagonal grid can be improved beyond their hexagonal distance by using safely connected non-adjacent moves, as we shall see shortly in Section 2.3.1.

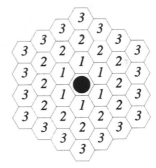

Figure 2.4. Rectangular and hexagonal distances based on adjacent moves.

2.2 Chains

A *chain* of pieces is a maximal set of connected pieces of the same color. Any piece within the set can be reached by any other piece though a series of adjacent moves that do not leave the set. Figure 2.5 illustrates the board position shown in Figure 1.2 with chains labeled.

Since it is impossible to separate adjacent pieces of the same color, each chain may be considered a single cohesive unit. This improves the complexity of Hex board analysis by reducing the number of elements to be examined from the total number of pieces on the board to the total number of chains on the board. For instance, since chain *a* touches both of White's edges in Figure 2.5 it is obvious from observing a single chain that White has won the game. For the purposes of Hex board analysis, singleton pieces and board edges are considered to be special cases of chains.

Chains are labeled with lowercase bold italic characters *a..z*. A set of chains is denoted by round brackets (*a, b*...) or by a single uppercase bold italic character (*C*) for convenience.

Empty board positions are labeled by lowercase italic characters in the range *p..z*. A set of empty board positions is denoted by curly brackets {*p, q*...} or by a single uppercase

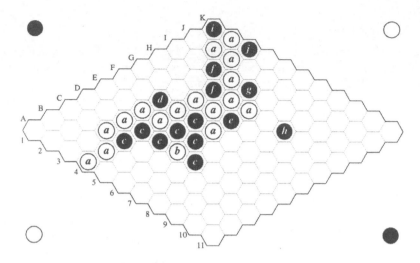

Figure 2.5. Pieces labeled by chain.

italic character $\{E\}$ for convenience. The set of empty points required by a connection is called the connection's *empty point set*.

2.3 Connectivity

The concept of connectivity is central to Hex. Two pieces or chains are *n-connected* if they can be joined by an unbeatable connection in n moves when considered in isolation. A connection is described as *safe* if $n=0$. All other connections are *unsafe*, with higher values of n indicating weaker connections. The isolation clause is necessary as two otherwise safe connections may overlap at some point, making them both unsafe when considered in combination. This definition of Hex connectivity is illustrated in Figure 2.6.

Figure 2.6(i) shows a pair of 1-connected pieces separated by the point p. White requires one move to complete a safe connection between a and b. The empty point set for this configuration is $\{p\}$.

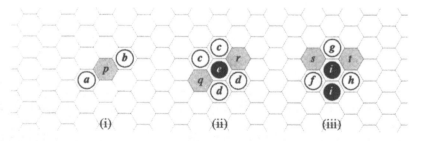

Figure 2.6. Hex connectivity: 1-connection (*a-b*), 0-connection (*c-d*), and 2-connection (*f-h*).

The two chains c and d in Figure 2.6(ii) are 0-connected as Black has no sequence of moves that can separate them; if Black plays at q or r, then White can reply by moving at the other empty point to complete the connection. The set of points $\{q, r\}$ forms the empty point set of this connection.

The concept of *cumulative connectivity* is illustrated by Figure 2.6(iii). Piece f is 1-connected with piece g and g is 1-connected with piece h, so the total connectivity between f and h is two as these lie in series. That is, White would require two moves to secure a safe connection between f and h.

More involved definitions of connectivity on the hexagonal grid exist [Wang and Bhattacharya 1997], but for the purpose of Hex board analysis *moves-to-connection* is the most appropriate due to the specific requirements of the game.

2.3.1 Bridges

Two pieces can be non-adjacent and still safely connected if they form the pattern illustrated in Figure 2.7. This configuration is sometimes called the *2-bridge* [Boll 1994] and here is described simply as a *bridge* to avoid confusion with the *n*-connectivity nomenclature (the 2-bridge is 0-connected). Schensted and Titus describe this structure as a Two-Way Stretch or Double Stretch, but then go on to use these terms to describe any fork or connection based upon two alternative links [1975].

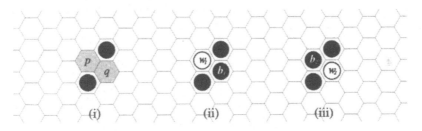

Figure 2.7. The bridge pattern and futile attempts to block it.

Empty points p and q form a *dual* that ensures the bridge's safety. If White plays at point p with move w_1, then Black can complete the connection by moving at the remaining empty point q with b_1 as shown in Figure 2.7(ii). Alternatively if White moves at point q with w_2, then Black completes the connection with move b_2 as shown in Figure 2.7(iii). The bridge connection is safe when considered in isolation.

To break this connection, the opponent must occupy one of the empty points p or q with a threatening move that simultaneously requires a reply elsewhere, distracting the player from defending the connection for at least one turn. This is difficult to achieve without careful planning and is discussed in more detail in Section 6.3. The bridge is not as secure as an adjacent connection but is the next best thing, and can be considered safe for the general case.

Figure 2.8 shows how *links* can be drawn to explicitly state the connections between pieces. The Figure on the left shows adjacent pieces linked through shared edges. The

Figure on the right shows a bridge linked across the two adjacent empty points through which it passes. Explicit linkage is useful for analyzing both connections between specific piece patterns and complete board positions.

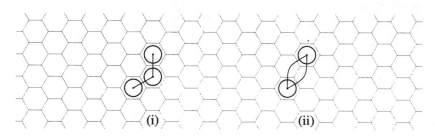

Figure 2.8. Links show connections between pieces explicitly.

Bridges allow faster spread of connections across the board, at the cost of a slightly less safe connection. Figure 2.9 shows the improved connectivity distances from a central piece using bridges in addition to adjacent moves. Compare these connectivity measurements to those using adjacent moves only (Figure 2.4). A 3-step link is shown to indicate how the expanded measurements are derived. Even greater connectivity distances across an empty board can be achieved by safely combining steps from chains, rather than just individual pieces. This is explained in Chapter 4.

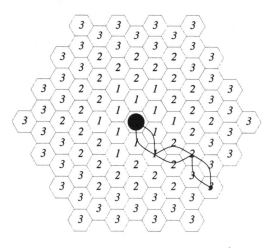

Figure 2.9. Expanded connectivity using bridges.

To further explore the nature of connectivity in Hex we will now examine some variants of the game. A deeper look at the hexagonal grid reveals why it is the optimal choice for connectivity games of this class.

2.3.2 Hex and the Game of Y

The game of Y is similar to Hex except that it is played on a triangular field of hexagons, and the winning condition is to complete a chain of pieces connecting all three sides. The two games are very closely related; Y was one of the games developed by Craige Schensted and Charles Titus after studying Hex in 1953. A good analysis of Y including discussions of key points of strategy can be found in *Mudcrack Y & Poly-Y* [1975]. For a review of this book see [Mallett 1998].

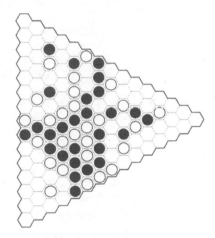

Figure 2.10. A game of Y won by Black.

A completed game of Y is shown in Figure 2.10. Black has won the game as they have formed a chain of Black pieces connecting all three sides of the board. The 15-per-side Y board provides 120 hexagons and is the closest match to the standard 11x11 Hex board with provides 121 hexagons.

Notice that goals are not marked on the Y board. This is because each edge does not have a specific color, and can be connected by either or both players. Like Hex, it is not possible to tie a game of Y.

The Hex board can be created from a Y board with pieces placed so as to create colored edges, as shown in Figure 2.11. Here an 11x11 Hex board is constructed from a 21-per-side Y board. Since Black has already connected the left and upper right sides, they must now make a connection between their row of pieces on the upper left and the lower right edge of the board. White must make a bottom left - top right connection. This is exactly the game of Hex, except for the complication of the swap option. However the chance of this formation actually occurring in a game of Y are extremely remote unless both players actively work towards this goal.

Conversely, the Y board can be constructed from a Hex board by delimiting a triangular region that does not touch any edges. This is a less direct conversion, as the rules must be modified to include the stipulation that players must connect the three sides of the

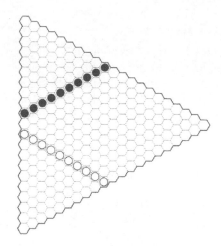

Figure 2.11. The Hex board is a subset of the Y board.

marked region. This is shown in Figure 2.12, where a 15-per-side Y board is created from a 17x17 Hex board.

Schensted and Titus claim that Y is the more interesting of the two games as Hex can be viewed as a subset of Y without the modification of rules [1975]. This claim is debatable—instead each game promotes different and specialized strategies. For instance:

• *the concept of playing to an obtuse corner is less relevant to Y than to Hex, where it is of the utmost importance, and*

• *due to the directed nature of Hex, strategies relative to one edge apply to the player's opposite edge when rotated 180 degrees about the central point, but do not apply relative to either of the opponent's edges. Any strategy in Y applies equally to all edges of the board.*

Note that the rules of Hex may be formulated as a subset of the rules of Y, but not vice versa.

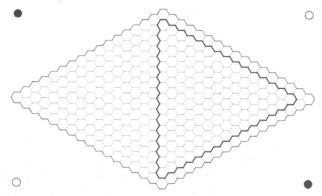

Figure 2.12. The Y board is a subset of the Hex board.

One obvious difference between the two types of board is that the Hex board contains two obtuse and two acute corners, whereas the Y board contains only acute corners. This means that points towards the middle of the Y board are always considerably further from the acute corner points than they are from the nearest edge. Schensted and Titus addressed this drawback by redesigning the Y board to contain three symmetrically spaced pentagonal cells near the middle of the board, each in line with a corner. The reduction of one edge for each of these cells means that the corner points are one step closer then they would have been on a strictly hexagonal board. The intention of Schensted and Titus was to make the modified Y board approximate a hemispherical mapping (where center-to-edge and center-to-corner distances would be as close as possible) without adversely affecting play. To achieve this they also slightly distorted the board's grid to give it a rounded appearance.

This raises the intriguing question of Hex played on non-hexagonal tilings. Note that such variants are actually games distinct from Hex, not just modified board representations as described in Section 1.7.

2.3.3 Other Tilings of the Plane

Let's consider a Hex variant played on the square grid called Square, for lack of a better name. The rules are the same as for Hex, only the playing surface has been changed to an 11x11 grid of squares. The only real difference between the two games is that cells in Hex have six adjacent neighbors, whereas cells in Square only have four adjacent neighbors.

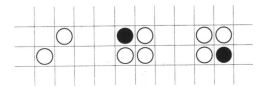

Figure 2.13. The bridge is less effective on the square grid.

This makes the bridge formation less effective, as shown in Figure 2.13. It is still safely connected but improves the connection by only one coordinate in each direction. As White is only really interested in horizontal movement the bridge does not give much additional benefit.

Unlike Hex, a game of Square is not guaranteed to result in a win for either player and may be *deadlocked* or tied. Figure 2.14 shows a deadlocked game that cannot be won by either player. The fact that it's deadlocked can be demonstrated beyond doubt by filling in the empty points as shown in Figure 2.15. Only one deadlock is necessary to ruin a game.

Why do deadlocks to occur in Square but not in Hex? This can be explained by examining the number of edges entering corner vertices on the board as shown in Figure 2.16. Let's call this number the *degree* of a vertex, notated as n-vertex.

Three cells meet at 3-vertices as on the hexagonal grid. When these cells are occupied there can be at most two chains of pieces collected around the vertex, as shown in the leftmost Figure.

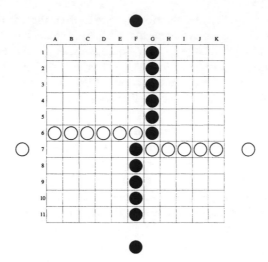

Figure 2.14. A deadlocked game of Square.

Four cells meet at 4-vertices as on the square grid. When these cells are occupied the formation of four disjoint chains is possible, as shown in the second part of Figure 2.16. The opposing chains of each player are not connected as adjacency is defined across shared edges, not corners. To complete their connection each player must maneuver around their opponent's pieces and connect via adjacent moves. This involves traversing additional 4-vertices which may themselves be deadlocked, and so on. The same holds for vertices of

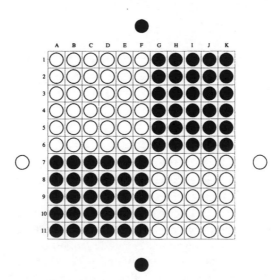

Figure 2.15. A full Square board does not guarantee a win.

Figure 2.16. Deadlock conditions exist for vertices of degree 4 or more.

degree 5 or more. *Any point at which more than two disjoint chains may occur around a vertex is susceptible to deadlock.*

For these reasons Square is not a very interesting game. The equivalent game Triangle played on a rhombus tessellated by triangles is even less interesting, as noticed by David Book [1998]. No analogy of the bridge exists on the Triangle board and all games would be doomed to deadlock; six disjoint chains are possible around each 6-vertex, and the greater the vertex degree the more likely it is for players to deadlock.

A tiling of the plane is *regular* if the tiling unit is a single regular polygon repeated periodically across the plane as shown in Figure 2.17. This has the attractive property of uniformity across the plane, but if deadlocks are to be avoided then we are limited to the hexagonal tiling as it is the only regular tiling with vertices of degree 3.

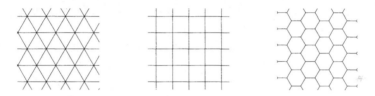

Figure 2.17. The three regular tessellations of the plane.

A richer variety of tessellations is available in the *semi-regular* tiling group, which involves the periodic tiling of two or more regular polygons. There are eight semi-regular tilings of the plane, two of which are shown in Figure 2.18. The tiling on the left is described as a 4.8.8 tiling and the one on the right as a 4.6.12 tiling based on the number of edges per tiling unit around each vertex. Both of these tilings satisfy the non-deadlock condition in that all vertices are of degree 3.

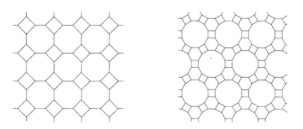

Figure 2.18. Semi-regular tessellations of the plane 4.8.8 and 4.6.12.

As a cell has at most as many connections as it has edges, tiles with more edges have a greater connective potential. For instance, the square tiles within the 4.8.8 tiling may be seen as having only half the connective potential of the octagonal tiles. This imbalance has a serious impact on game strategy for semi-regular boards and it is preferable to avoid tessellations composed of tiles with a greatly disparate number of edges.

Irregular tilings of the plane based on irregular polygons allow still wider variety. Figure 2.19 shows an irregular tiling proposed by David Book [1998] that provides an interesting game. The discrepancy between the number of edges per tile (between four and eight) is less pronounced on this board as the stronger tiles are more widely spaced than in the two semi-regular tilings shown, and interspersed with intermediate tiles. This tessellation does not contain any potential deadlocks.

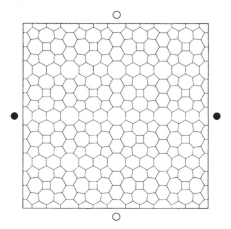

Figure 2.19. This irregular tessellation makes an interesting board.

Book also proposes that Hex may be played on any map of sufficient size, for example a map of the contiguous United States (Figure 2.20). White tries to connect the Pacific and Atlantic oceans while Black tries to connect Mexico with Canada.

Black has an easy win with first move by stepping through California, Utah (a bridge move), Wyoming, and South Dakota as shown. A possible deadlock exists where four states (Utah, Colorado, Arizona, and New Mexico) meet at a common vertex to the bottom left of the board. It can be seen that this map does not provide a very interesting game. More finely grained maps such as those at the county level may be necessary to allow richer strategy. This example is particularly apt, as the idea that became Hex occurred to Piet Hein while he was contemplating the famous four-color theorem of topology [Gardner 1959].

Another tiling of interest is the Voronoi tessellation of a set of points in the plane. The Voronoi diagram of a set of n points is a partitioning of the plane in n convex polygons, such that each polygon contains exactly one point and every point within each polygon is closer to its center than to the center of any other polygon [Weisstein 1999]. The Voronoi diagram of a point distribution has many desirable qualities: the cells it produces are con-

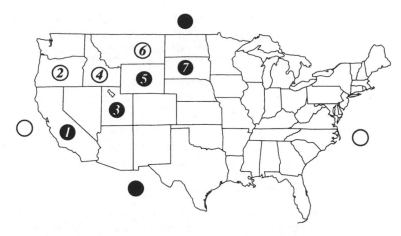

Figure 2.20. Black can force a win on the map of the United States.

vex and tend to be reasonably regular. Efficient algorithms exist for calculating the Voronoi diagram for a given set of points [O'Rourke 1998]. A Hex board based on the Voronoi diagram of a set of randomly distributed points is shown in Figure 2.21. This game is called Vortex.

Point distributions can be constrained or the subsequent diagram modified to ensure that no deadlocks occur and tiles are reasonably uniform in size, shape and distribution. Such a board is described as *well-behaved*. All of the vertices in the board illustrated are of degree 3, though this is not obvious when displayed at low resolution. For an additional artistic touch the distribution may be postprocessed to encourage more regular angles and edge lengths within each tile.

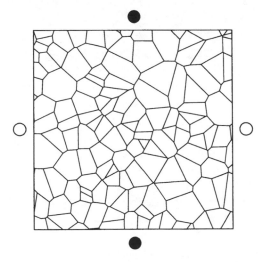

Figure 2.21. Vortex board formed by a Voronoi tessellation.

One of the interesting points of Vortex is that a unique board may be generated for every game. It's up to the players to analyze this structure and use it their advantage. Such shifting boards truly test a player's knowledge of the game and their ability to apply known strategies to unique situations.

The flexibility of the Voronoi diagram may be exploited to fit the game to shapes other than the rhombus. For instance a Circular Vortex board can be generated by dividing a circle's perimeter into four equal regions, assigning opposite regions to each player, and filling the interior with a well-behaved point distribution. Circular Vortex eliminates the corners of the Hex board. The game is still directional in nature, as each player has denoted goals they must connect, but significantly different edge handling strategies are required.

2.3.4 Tiling Three-Dimensional Surfaces

Expanding the discussion of planar tilings to include tessellated surfaces of three-dimensional objects allows even more variety in playing boards. A particularly attractive tessellation is the 5:6:6 tiling that constitutes Buckminsterfullerene, or the Bucky Ball, shown in Figure 2.22. This shape is commonly recognized as the pattern on a soccer ball.

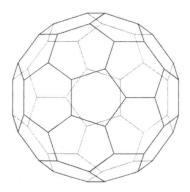

Figure 2.22. Buckminsterfullerene.

Buckminsterfullerene is a regular polyhedron (each face is a regular polygon). It has the desirable qualities of containing only 3-vertices, and being composed of tiles of similar connective potential. Its only drawback as a playing board, however, is a major one: it has only 32 faces and hence would not allow a very interesting game.

As the tiling is periodic in all directions around the surface there are no edges or other distinguishing features. The game is not *oriented* in that it has no intrinsic point of reference that can be used to define objectives for each player.

The objectives of the game must therefore come from the players' pieces themselves, which changes the nature of play considerably. The most elegant winning condition for this type of game is the *Surrounded Cell* rule as follows:

•   *the player to first surround at least one cell (empty or occupied by the opponent) with a continuous chain of their pieces wins.*

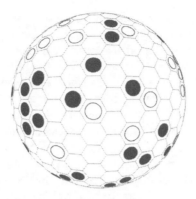

Figure 2.23. Hex on the sphere?

This replaces the previous winning condition for Hex stated in Section 1.2 but otherwise the game is similar. The swap option may still be employed to discourage the first player from opening on one of the more powerful (hexagonal) cells.

For a larger playing area we must look to semi-regular or irregular tilings of the three-dimensional surface. Can a sphere be tessellated with hexagons as shown in Figure 2.23 to provide a periodic analog of the Hex board? Unfortunately it cannot. Even if irregular hexagons were used the board would only look "correct" from the angle shown and its antipodean pole. If the object shown were made into a model, it would look increasingly distorted toward the equator as it was rotated in the hands.

The reason for this is shown in Figure 2.24. On the left is an area tiled by regular hexagons. The six edges of the central hexagon result in the area's six corners, marked *v*. Notice that these corner cells have three exposed edges, whereas non-corner cells on the outer layer only have two exposed edges.

This tiling may be mapped to each of two hemispheres making up a sphere, but to ensure a seamless join the six corner cells of each must be changed to pentagons as shown on the right of Figure 2.24. This means that the resulting sphere must contain at least

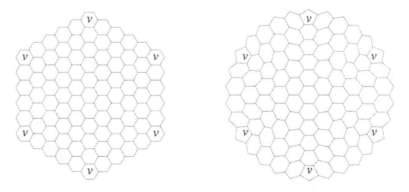

Figure 2.24. Hexagonal grid modified to wrap around a sphere.

twelve pentagons (six for each hemisphere) regardless of its size. The presence of these flaws orients the sphere in space and makes it a less attractive playing surface.

It's possible to maintain the illusion of a spherical mapping by showing to the player a view of the expected distribution of hexagons, and hiding the irregularities around the edges where the projected view vanishes. This is not a true spherical mapping; it is much the same effect as a fish-eye lens moving over a periodically repeating game on the plane, such as occurs in Modulo Hex (Hex that wraps around rather than stopping at edges, described in Appendix E.1). All pieces will always be visible, and the view of the board can be scrolled smoothly in any direction. Of the suggested winning conditions for Modulo Hex only the Surrounded Cell rule applies when played on the hemisphere.

A well-behaved Voronoi tessellation of a point distribution across the surface of a sphere would provide a suitable playing surface. It would be unoriented and viewable from any angle.

2.3.5 Three-Player Hex

As the largest prime factor of the number of sides of a hexagon is three, the possibility of play between three opponents seems natural. This is explored using a game called Three-Player Metahex, so named due to the fact that it's played by three opponents on an area composed of a large hexagon tessellated by a number of smaller hexagons. Each opponent owns two directly opposed edges and tries to connect them with a continuous chain of their pieces.

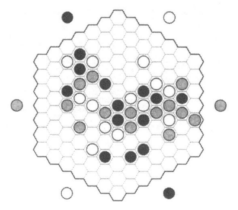

Figure 2.25. A game of 3-Metahex in progress.

An interesting feature of the Metahex board is that it contains only obtuse corners. A game in progress is shown in Figure 2.25.

Although it is not immediately obvious, this game is already deadlocked and no player can win. Figure 2.26 shows that a game can be tied even if all positions are occupied. Again, a single deadlocked vertex ruins the game.

Our previous observation that 3-vertices cannot be deadlocked must now be modified

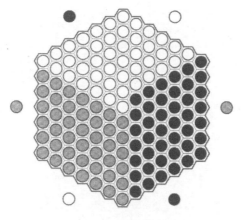

Figure 2.26. Three-Player Metahex can be tied.

to handle the three-player case. This is illustrated in Figure 2.27 where it is shown that the three players may indeed combine to form a deadlock around a 3-vertex.

The previously stated rule that any point at which more than two disjoint chains may occur around a vertex is susceptible to deadlock still holds for the three-player case. However, it must now be realized that the third player may complete the deadlock by isolating all three pieces around a 3-vertex.

Figure 2.27. The three-player deadlock condition.

In addition, Figure 2.28 shows that bridges are not safe connections for three-player games on the hexagonal grid. Consider Black's bridge: if White intrudes at point p, then Gray can complete the block by playing at its dual point q. This is not to the advantage of any of the players, however, as two deadlocked vertices have now been created.

This illustrates how each player is effectively pitted against two opponents in Three-Player Metahex and has next to no chance of victory. Connections must now have *three or more alternative paths* to be safe. This is extremely difficult to achieve as both opponents have the chance to block before the player's next turn.

Three-Player Metahex is of limited interest as nearly all games will end in a tie among the three players. One solution to this problem is the *McCarthy Revenge* rule proposed by Straffin: as soon as it is no longer possible for a player to connect their sides they are eliminated from the game, resulting in a two-player game [1985].

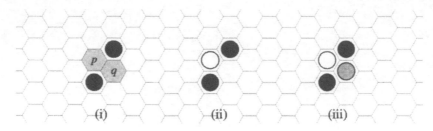

Figure 2.28. Bridges are not safe connections in three-player games.

Summary

Pieces at adjacent points on the Hex board are described as touching. Sets of touching pieces are called chains.

The connectivity between a pair of points or chains is the number of moves required to guarantee the connection. Connections of order zero are safe connections. Bridges are non-adjacent but safely connected.

Hex and the game of Y are extremely closely related. Most strategies that apply to one game will apply to the other.

The hexagonal grid is the optimal choice of playing surface because it:

•   *is uniform in size, shape and point distribution,*

•   *is the regular tiling with the greatest number of neighbors and hence connective potential,*

•   *allows bridge moves, and*

•   *does not allow deadlocks.*

Other suitable grids such as the 4:5:6:7:8 tiling and well-behaved Voronoi tessellation allow interesting variations on the existing game.

The Surrounded Cell winning condition allows play on unoriented three-dimensional surfaces. The most elegant regular polyhedron for this purpose, Buckminsterfullerene, unfortunately provides too small a playing surface for an interesting game. Hemispherical mappings and Voronoi diagrams on the sphere appear more suitable.

Three-player games played on the hexagonal grid are of limited interest.

Strategy I: Basic

This chapter introduces some basic points of strategy that do not require a deep understanding of Hex. More advanced points of strategy are discussed in Chapter 6 after some further concepts regarding piece connectivity are discussed in Chapters 4 and 5.

The points of basic strategy fall into three broad categories: structural, positional, and general.

3.1 Structural Development

Structural development involves the exact placement of pieces on the board. It is central to the game of Hex, which is won or lost depending on the connectivity between pieces, and is the most important aspect of basic strategy.

3.1.1 Expand by Bridges

As bridges (defined in Section 2.3.1) cover twice the board distance as an adjacent move in the same direction, they allow a player to spread a safe connection across the board twice as fast as a player using adjacent moves. Although they can conceivably be broken by the intrusion of a nearby play that requires an immediate reply, bridges are in most cases as good as an adjacent connection, and are an important element of Hex strategy. They are the cornerstones upon which a player's connective structure is usually based.

Consider the example shown in Figures 3.1 and 3.2. Four adjacent moves from the central hexagon (Figure 3.1) cover only half the distance of four bridge moves from the central hexagon (Figure 3.2).

3.1.2 Maximize Connectivity, Minimize the Opponent's

The strength of a player's board position is based on the connectivity of their pieces across the board. This can be determined by estimating how many steps (pieces to be placed) the player requires to create an unbeatable chain between their goals. Adjacent pieces and

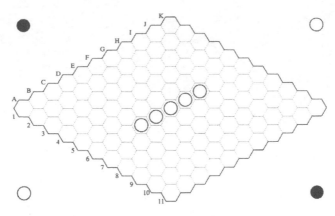

Figure 3.1. Adjacent moves from the central hexagon.

bridges count as a single step in this calculation. Best-case connectivity is a win for the player (zero steps required), worst-case connectivity is a loss (opponent needs zero steps to complete their chain), and while the outcome of the game is undecided, connectivity lies somewhere between these extremes.

The fundamental law of Hex strategy is that a player's *position is only as good as the weakest link in their best connection across the board.* With each move, the player should attempt to improve their weakest link or exploit their opponent's weakest link. Any move that achieves both of these objectives at once is a strong move.

For instance, the board position shown in Figure 3.3 may appear to favour Black, but either player can win with the next move if it is their turn. Point p is obviously the weakest link in Black's strongest connection across the board. Moving there would link pieces c and d to achieve a spanning connection of Black bridges across the board. Not so obviously, the point p is also the weak link in White's strongest connection across the board,

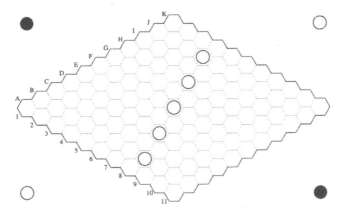

Figure 3.2. Bridge moves from the central hexagon.

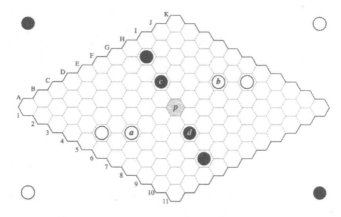

Figure 3.3. Point *p* is the weakest link in both players' strongest connection.

and moving there would link pieces *a* and *b*, giving White the win This key point is the weak link that each player must try to secure and the opponent to exploit.

This is another reason why F6 is such a strong opening move: the first player's weakest link at the outset of the game (edge-to-edge) is halved. If the second player replies with a move near the center they also improve their weakest link, but for reasons explained in Section 3.1.3, the first player can evade such close-in play and their dominance is not threatened.

Towards the end of a game it is common for both players' connections to be well defined across most of the board, but somewhat ambiguous in the region where the opposing connections cross over. This uncertain region is invariably each player's weakest link in their best connection, and must be resolved as soon as possible. Whoever wins this battle will win the game.

A common danger in Hex is the tendency to "win the battle but lose the war." It may be satisfying to complete a hard-fought connection in one area of the board, but this is of little use if it does not improve the player's overall position. In fact it may have given the opponent the opportunity to place strategic pieces during their local defense that have a serious impact on the rest of the game.

3.1.3 Start Blocking at a Distance

In order to block an opponent's connection it's tempting to play close to the leading piece of their attack. However, the topology of the hexagonal board and expansion of bridges make it easy for the opponent's pieces to flow around close blocks. It's generally more effective to prepare blocking moves some distance from the advancing connection.

Figure 3.4 illustrates the importance of blocking at a distance. Imagine that White's chain of pieces is safely connected to the top right edge, and is now threatening to advance towards the bottom left edge.

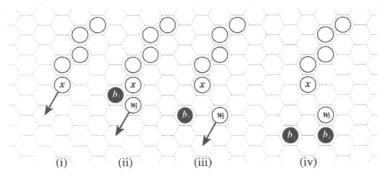

Figure 3.4. Three attempts to block White's advancing connection. The arrow indicates the
direction of White's attack.

If Black attempts to deflect White's advance piece *x* with the adjacent block b_1 then
White can flow around unimpeded with move w_1. Even a close non-adjacent block such as
b_2 is ineffective, as White can use bridge *w2* to once again move around the block and
continue the attack. Reply b_3 to White's advancing piece *x* proves to be a more formidable
block. White's attempt to step around the block with bridge w_3 can be successfully countered by move b_4. The defensive block b_3 is called the *classic defense* [Boll 1994].

The key to this strategic point is that blocks made further away from the attacking
piece have a move or two before the advancing head reaches them. They are therefore
more flexible, and can adapt to variations in attack. Playing too close to the attacking piece
commits the defender to a certain line of defense, which may not be the optimal one. The
defending player should also keep in mind that bridges extend the opponent's range to
twice that of adjacent moves, so a solid defense must also cover this extended threat.

3.2 Positional Play

Positional play is less important than structural development, but often helps the player
select an optimal move from a set of otherwise equally good alternatives. The following
points of strategy should be regarded as rules of thumb rather than hard-and-fast rules.

3.2.1 Occupy the Short Diagonal

Pieces placed along the short diagonal that connects the board's two obtuse corners are in
a strong position. They are well placed to connect to the nearest home edge and to block
the opponent's pieces from their home edges. The central point along the short diagonal
(F6) is the strongest opening move, and gives the first player a winning advantage. For
games on even-sided boards, the two central points along the short diagonal are equally
strong opening moves.

Points along the short diagonal are equally strongly connected to the two nearest edges
(one edge belonging to each player) as illustrated in Figure 3.5. This forms a shared borderline that is advantageous for both players to occupy, especially since a move on a short

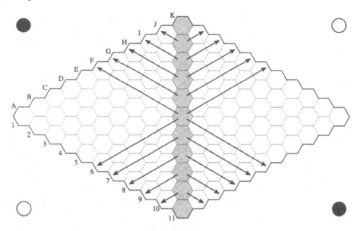

Figure 3.5. Strong territorial influences from points along the short diagonal.

diagonal point prevents the opponent from taking that point themselves. F6, the midpoint along the diagonal, is special in that it exerts strong influence to all four edges, hence its value as a strong opening move.

Playing to either side of the short diagonal is usually not as strong as a move on the diagonal itself. A move towards the player's edge is defensive and may be readily connected to the edge, but faces the danger of being cut off if the opponent claims the diagonal in that area. Alternatively, a move towards the opponent's edge may be overly aggressive and runs the risk of being isolated and effectively removed from the game, wasting a move.

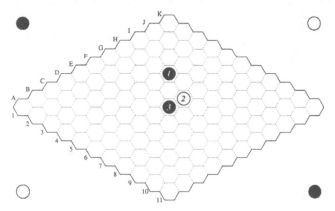

Figure 3.6. A position where White has played a weak off-diagonal move.

The board position shown in Figure 3.6 is a case in point. White opens with *1* H4 (far too strong an opening move!) which Black swaps to give them first move. White then replies with *2* at G6. This is a very weak move; it does not block piece *1* effectively and is in no-man's land just off the diagonal. Black's solid move *3* F6 is on the diagonal and threatens to link to piece *1*. Black now commands the center, has almost guaranteed an easy connection to the top, and is in a commanding position.

Given a choice between otherwise equally good moves, prefer moves along the short diagonal, and occupy the midpoint if possible.

3.2.2 Don't Play Too Close to Existing Pieces

Play during the early stages of the game (and also play into unclaimed board territory) is characterized by moves that are well-spaced from a player's own pieces, minimizing the largest gap between them, and also well-spaced from opponent's pieces to allow room to block if necessary. This forms a sparse framework of connections which fills out and strengthens as the game progresses. Play gets closer to existing pieces (more aggressive) as connections become more well-defined.

Figure 3.7 shows the first six moves in a game between two intermediate players, illustrating how the early moves in a game are usually spread out from each other to form sparse connections that become firmer as the game progresses. Although only six moves have been played, the connective framework for the rest of the game has already been established. Presumably this game was played without the swap option, accounting for the strong opening move F6.

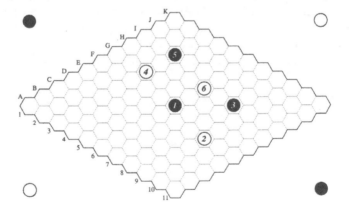

Figure 3.7. Early moves are usually spread out from existing pieces.

This point of strategy is more of an observation or rule of thumb. In many cases it is advantageous to reply close to opponent's initial moves; for instance, if their move is one point away from the diagonal it may be prudent to play in the adjacent point along the diagonal. Figure 3.6 illustrates a situation where close-in play is appropriate from the early stages of the game. If the opponent makes a weak move, the player should punish it. Experienced players often play "close in" from the game's outset in order to intimidate opponents with this aggressive style of play.

3.2.3 Respond to the Opponent's Last Move

As explained in the last section it is usually good to keep some distance from existing pieces. However, it can also be unwise not to respond to the opponent's last move, which may require playing near the piece just placed. This has a number of potential advantages:

- *it may interfere with the opponent's strategy in that region,*

- *it avoids the opponent gaining free territory unhindered,*

- *it reduces the chance of the opponent developing momentum in that region (see Section 6.3), and*

- *it may foil a future trap if the opponent's move was intended as a ladder escape (Section 7.3).*

There is little danger of the player focusing too strongly on their connection and ignoring their opponent's; by definition either connection is severed when the other is completed. However, the player should remember that their opponent is playing to a strategy, and ignoring their most recent move may prove costly. Examining the opponent's last move may also provide some insight into their overall strategy.

As with the previous point of strategy this is more of a rule of thumb, and if a better move on another part of the board is evident then that move should be taken. This strategy (or any other) should not be followed blindly, and the player should be wary of being manipulated into a series of forced replies and losing the momentum of the game. However, playing near the opponent's last move crowds them in and exerts some pressure on their attack.

3.3 General Strategies

The following points of general strategy don't fit into the context of structural or positional play. They describe overall strategies that should be kept in mind over the course of a game, and should influence play where appropriate.

3.3.1 Play Defensively

If a player completes a safe connection between their sides, then the opponent is prevented from completing theirs. Conversely, if the opponent is prevented from completing a safe connection, then the player must be able to complete theirs and win. This means that good defensive play will often win a game as quickly as strong aggressive play. A most important strategy is to *play defensively unless there is a good reason not to do so.*

Aggressive play should generally be curtailed until the later stages of the game, or when a player wants to shake up the game to create an opening or otherwise intimidate an opponent. If a player is losing, aggressive moves may change the state of play.

It is possible to play *too* defensively and get caught up in a minor skirmish while losing touch with the overall position. The player must learn to see beyond the local connection and instead understand its part in the global connection.

Before making a move, players should always ask themselves *"what is the most damaging reply the opponent can make?"* This simple tip is useful for exposing moves that superficially look sound but have dire consequences. If a killer reply can be found to one move, often the same reply immediately repudiates a number of similar moves that can be removed from the potential list of moves without further ado. Levy describes this mechanism in terms of computer game-players as the *killer heuristic* [1983].

There is a certain degree of balance and self-containment within Hex:

- *one player must win, one player must lose,*

- *the existence of a player's connection precludes the existence of the opponent's connection, and*

- *good defense and good attack are equivalent.*

These observations reflect the eastern concept of Yin and Yang, the harmony achieved by the balance of opposed forces of aggression and passivity. A game of Hex is won or lost when this balance is overthrown.

3.3.2 Foil the Opponent

If no obviously good move presents itself a good rule of thumb is to *play where the opponent would most like to move next turn*. This strategy of *foiling* the opponent's next move is sound, as weakening the opponent's position will always improve the player's position.

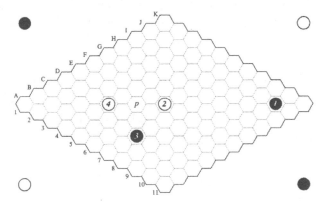

Figure 3.8. Move where your opponent would most like to move next turn.

For example consider the board situation illustrated in Figure 3.8. It's early in the game and neither player has a well-developed connection or obvious strategy yet, so where should Black move next?

In the absence of any better choices Black would be wise to move at point *p*. White would dearly like to play there next turn to connect their central piece *2* to piece *4* and therefore to the lower left edge, a very commanding position. This example also highlights the importance of responding to the opponent's last move, as discussed in Section 3.2.3.

The board situation shown earlier in Figure 3.3 is another case in point. If Black plays at *p* next move they win the game. If it were White's turn to move, they would be wise to play at *p* themselves to postpone defeat, forcing Black to find another way to make the connection.

3.3.3 Don't Resign Too Early

This point sounds obvious but is more important in Hex than in most other games. It is Hex etiquette to concede the game once its outcome becomes obvious to both players. Due to the *deterministic* nature of Hex (see Section 1.4.3) there will be no random intervention that may save a player from a losing position.

So when exactly is a game over? The rules state that the game is won when either player completes a connected chain between their sides of the board. However, this winning condition is somewhat redundant as continuing to move once a player has an unbeatable connection across the board is clearly a waste of time. In practice the game finishes with the resignation of the player on the losing end of the unbeatable connection, once they are satisfied that their opponent has also spotted the connection and is able to maintain it. Hence the exact moment of the game's finish largely depends on the two players' level of skill.

For a good example of this point see Sample Game 11.3. White fumbles a guaranteed win and resigns in frustration without noticing a second winning move that has been on the board for a number of turns. Although Black may have been tempted to resign immediately upon noticing the losing situation, they persevered until it became apparent that White had not noticed either winning play. Black was eventually rewarded with the win.

The rule applies mostly to novice players. The chances that an experienced player will fail to see a winning situation or make a disastrously bad move are rather slim—though it has been known to happen!

3.3.4 Assume the Worst

Sometimes a particular move will appear too good to resist. This may be the culmination of a clever line of play that has been developed for some time, leading to a trap for the opponent. Many players will make this move even if a spoiling play is available to the opponent. There is the temptation to assume that the opponent will fail to see the spoiling move and fall for the trap instead. However, this is a very bad assumption to make!

A great play with a minor flaw is not as strong as a mediocre but solid play with no flaws. This is a variation of the point discussed in Section 3.1.2.

It's always wise to *base play on the worst-case scenario*, even against less experienced opponents.

3.4 Applying Basic Strategy

Players should make the structural development of pieces a top priority. If several equally good moves are available for a given turn, then positional aspects may be useful in deciding which of those moves is best. Points of general strategy should be kept in mind throughout the game and influence play where appropriate.

In general a player should:

- *identify the most likely winning line(s) based on their current board position, and try to set up at least one of these,*

- *identify threatening plays by their opponent and try to thwart them, and*

- *strengthen and expand their connection with good positional play.*

The first two points define exact plays to be made. The third point is less tangible and becomes more important in narrowing down the choice of moves. The opponent's connection is weakened if the player's connection is strengthened, by definition. Any move that combines all three points simultaneously is a good move indeed.

Above all: *play defensively if uncertain.*

Summary

Points of basic strategy fall into three broad categories: structural, positional and general. Structural and positional strategies are most important and often indicate precise moves as the optimal choice. General strategies are less tangible, and used more to influence the choice between otherwise equally good moves.

A player's position is only as good as the weakest link in their best connection across the board. Playing defensively is generally the most sound approach. If no good moves are obvious then it is a good idea to play where the opponent would most like to play next turn.

Groups, Steps, and Paths

Connectivity across the board is described in terms of groups (collections of intercon-nected chains), steps taken from groups, and paths forming connections between groups. These terms are defined precisely, and a path algebra capable of describing the overall connectivity of a given board position is introduced.

4.1 Groups

A *group* is a connected set of chains and empty points. Any chain or empty point within the group can be reached from any other by moving through a series of adjacent hexagons within the group. A group's connectivity is given by the worst of the minimum connectivities between individual pairs of chains within the group.

The general form of a group is:

> *n-group(**C** {S})*

where *n* is the connectivity of the group, *C* is the set of chains comprising the group, and *S* is the group's empty point set. If there are no members in *S* then the group collapses to a single chain.

Figure 4.1 shows a variety of groups. The links comprising each group's connections are shown directly below each example. Links comprised of paired arcs between two pieces indicate a bridge connection.

A *safe group* is a group in which every combination of chain-to-chain pairs is 0-connected. Safe groups are described in the format *<C {S}>* where *C* is the set of chains that comprise the group and *S* is the group's *empty point set*. Groups with at least one chain-to-chain pair that is not 0-connected are called *unsafe groups*.

The simplest form of safe group consists of a single chain as shown in Figure 4.1(i), and is a called a *singleton group*. Singleton groups have no associated empty point set and are described in the format *<c>* where *c* is the identifier of the single chain member.

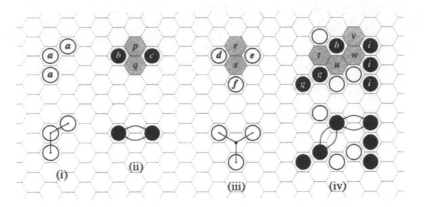

Figure 4.1. Examples of groups: singleton group, safe group, unsafe group, and another safe group.

Non-singleton groups (groups consisting of two or more chains) must have at least one empty point separating them, or they would converge to a single chain. For instance, chains b and c in Figure 4.1(ii) form a non-singleton group, and $\{p, q\}$ is the empty point set separating them. This configuration is the bridge pattern introduced in Section 3.1.2; hence, this group is 0-connected and safe.

The set of chains (d, e, f) across the region $\{r, s\}$ shown in Figure 4.1(iii) is an unsafe group as chain pairs (d, f) and (e, f) are 1-connected. However $<(d, e)\{r, s\}>$ forms a safe group if chain f is excluded.

The group illustrated in Figure 4.1(iv) is safe as chain pairs (g, h) and (h, i) are 0-connected by bridge patterns, and chain pair (g, i) is 0-connected through the common chain h.

A safe group may consist of subgroups of unsafe connectivity. For instance, the 0-group$((g, h, i)\{t, u, v, w\})$ illustrated in Figure 4.1(iv) contains of the following subgroups:

0-group$((g)\{\})$

0-group$((h)\{\})$

0-group$((i)\{\})$

0-group$((g, h)\{t, u\})$

1-group$((g, h)\{t\})$

1-group$((g, h)\{u\})$

0-group$((h, i)\{v, w\})$

1-group$((h, i)\{v\})$

1-group$((h, i)\{w\})$

2-group$((g, i)\{u, w\})$

0-group$((g, h, i)\{t, u, v, w\})$

1-group$((g, h, i)\{t, u, v\})$

1-group$((g, h, i)\{t, u, w\})$

1-group$((g, h, i)\{t, v, w\})$

1-group$((g, h, i)\{u, v, w\})$

2-group$((g, h, i)\{t, v\})$

2-group$((g, h, i)\{t, w\})$

2-group$((g, h, i)\{u, v\})$

2-group$((g, h, i)\{u, w\})$

For the purposes of Hex board analysis, we are only concerned with safe groups and unsafe groups may be discarded. The importance of safe groups will be explained in more detail shortly.

Discarding unsafe groups also reduces the number of possible group combinations to be examined, substantially improving the efficiency of the path analysis algorithm described in Chapter 8. For instance, the above list of subgroups derived from Figure 4.1(iv) reduces to a much smaller set of safe groups:

$<g>$

$<h>$

$<i>$

$<(g, h)\{t, u\}>$

$<(h, i)\{v, w\}>$

$<(g, h, i)\{t, u, v, w\}>$

Future use of the term *group* will refer to safe groups only.

A group's empty point set should be the smallest possible set for that group. If any point can be removed from the empty point set without making the group unsafe, then that smaller point set must be used instead. In other words, *empty point supersets are not permitted.* This is because any points outside the minimal set are superfluous and can only be to the disadvantage of the group, especially when trying to place the group on a crowded board. Thus point r is excluded from the connection links of Figure 4.1(iii).

Two groups are *disjoint* if they do not share any common chains and their empty point sets do not overlap at any point.

When describing groups, steps, and paths the following conventions will be used:

- *White cells: not part of the group.*

- *Shaded cells: members of the group, step or path in question.*

- *Dotted cells: pivot or vulnerable points within the empty point set.*

4.2 Steps

Just as a chain is as safely connected as a single piece, a safe group is *almost* as safely connected as a chain, and forms a solid foundation from which to build connections across the board. These connections are based on *steps* that connect a group to an empty point, then spread from that point outwards.

The general form of a step is:

$$n\text{-}step(<C \{S_1\}> \underline{p} \{S_2\})$$

where n is the connectivity of the step, $<C\{S_1\}>$ is the group from which the step is taken, $\underline{p}$ is the point to which the step is taken, and S_2 is the empty point set required by the step. The point $\underline{p}$ is underlined to indicate that it is the *terminal* or *pivot point* of the step. If no empty point set S_2 is shown, then the step is an adjacent step that does not require any intermediate empty points.

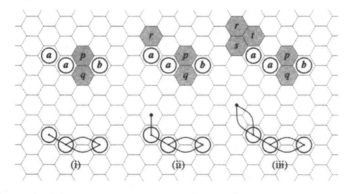

Figure 4.2. A group, an adjacent step, and a bridge step taken from the group.

The connectivity of the step equals the number of moves required by the player to safely connect the pivot point with the group. For example Figure 4.2 shows safe group $<(a, b)\{p, q\}>$ and two steps leading from it. Both of these steps are of connectivity 1, as each step requires a single move at the pivot point before the step becomes safely connected to the group.

1-step($<(a, b)\{p, q\}>\underline{r}$) shown in Figure 4.2(ii) is an adjacent step without an empty point set. Notice that the terminal point $\underline{r}$ is dotted in the lower diagrams to indicate that it's a pivot point. The step shown in Figure 4.2(iii) is 1-step($<(a, b)\{p, q\}> \underline{r} \{s, t\}$) and forms a bridge move from the group. A large number of additional adjacent and bridge steps may be taken from this group.

Steps are *commutative*, that is $n\text{-}step(<C\{S_1\}> \underline{p} \{S_2\})$ and $n\text{-}step(\underline{p} <C\{S_1\}> \{S_2\})$ are equivalent. This property becomes important for the efficient implementation of a path-finding algorithm (described in Chapter 8, Algorithmic Board Evaluation).

Notice that *no part of a step may overlap the group's own empty point set*. This is a fundamental rule that leads to the definition of two step operators that can be used to combine steps and advance the spread of connectivity across the board:

4.2.1 Step Consolidation ⊗

Two steps of connectivity 2 from the same group $<C \{S_1\}>$ that terminate at the same pivot point $\underline{p}$ can be *consolidated* if their empty point sets do not overlap each other or the group's own empty point set, to form a single step of connectivity 1. That is:

$$2\text{-}step_1(<C\{S_1\}> \underline{p} \{S_2\}) \ \otimes \ 2\text{-}step_2(<C\{S_1\}> \underline{p} \{S_3\}) \ \Rightarrow \ 1\text{-}step(<C\{S_1\}> \underline{p} \{S_4\})$$

$$\textit{iff } \{S_1\} \wedge \{S_2\} \wedge \{S_3\} = \{\}$$

$$\text{where: } \{S_4\} = \{S_2\} \vee \{S_3\}$$

Figure 4.3 shows 2-step($<a> \underline{p} \{q\}$) and 2-step($<a> \underline{p} \{r\}$) consolidating to form 1-step($<a> \underline{p} \{q, r\}$).

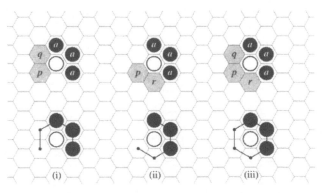

Figure 4.3. The consolidation of a pair of 2-connected steps between group $<a>$ and pivot point $\underline{p}$ to form a single 1-connected step.

The connectivity of the resulting path is one less than that of the original steps, reflecting the fact that the player has an alternative route if one of the component steps is blocked. This is why it is critical that the empty point sets are disjoint; if there were any point of overlap, then the opponent could play at this point and *intrude* on either or both steps, a step and the group, or all three at once.

Step connectivity cannot be improved beyond one since at least one move is required to occupy the pivot point; hence, it is invalid to consolidate two 1-steps. Steps of connectivity three or more may not be safely consolidated as they are inherently vulnerable. This is discussed in more detail in Section 4.3.5.

4.2.2 Step Extension ⊕

While step consolidation strengthens existing connections, the *step extension* operator facilitates the spread of connectivity across the board at the expense of weaker connection.

A step from group $<C\ \{S_1\}>$ to pivot point p can be *extended* by a further step to pivot point q if their empty point sets do not overlap each other or the group's own empty point set. That is:

$$n_1\text{-step}(<C\{S_1\}> p\ \{S_2\})\quad \oplus\quad n_2\text{-step}(p\ q\ \{S_3\})\quad \Rightarrow\quad n_3\text{-step}(<C\{S_1\}> q\ \{S_4\})$$

$$\textit{iff } \{S_1\}\wedge\{S_2\}\wedge\{S_3\} = \{\}$$

$$\text{where: } n_3 = n_1 + n_2$$
$$\{S_4\} = \{S_2\}\vee\{S_3\}\vee p$$

Figure 4.4 illustrates *step extension* from 1-step($<a> r\ \{p,\ q\}$) to 2-step($<a> u\ \{p,\ q,$ $r,\ s,\ t\}$). Again there is no overlap between any of the empty point sets. Notice that the pivot point r from which the additional step is taken is added to the extended step's empty point set, and its status as a pivot point is remembered. Pivot points are of struc-

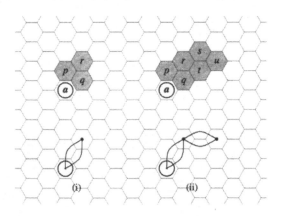

Figure 4.4. Step extension from 1-step($<a> r\ \{p,\ q\}$) to 2-step($<a> u\ \{p,\ q,\ r,\ s,\ t\}$).

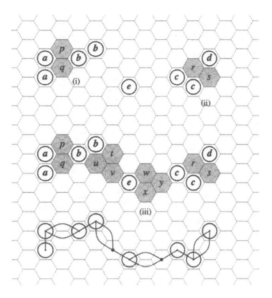

Figure 4.5. Two disjoint groups and a 2-path connecting them.

tural importance within the overall step; they are vulnerable points that should be singled out if the step is to be attacked or defended, and indicate points that allow connection in the fewest possible moves.

4.3 Paths

A *path* is a connection between a pair of disjoint safe groups described by a set of empty points and intermediate chains used as stepping stones between the groups.

The general form of a path is:

$$n\text{-}path(<C_1 \{S_1\}> <C_2\{S_2\}> C_3 \{S_3\})$$

where n is the path's connectivity, $<C_1 \{S_1\}>$ and $<C_2 \{S_2\}>$ are the two disjoint safe groups that form the *terminal groups* of the path, C_3 is the set of *intermediate chain set* of chains used as stepping stones between the terminal groups, and S_3 is the empty point set that describes the actual path between the groups. The intermediate chain set C_3 must not intersect with the chain set of either terminal group (C_1 and C_2), and the empty point set S_3 must not intersect with empty point set of either terminal group (S_1 and S_2). Note that C_3 is not a group, it is just a collection of chains.

Figure 4.5 shows two disjoint safe groups $<(a, b)\{p, q\}>$ and $<(c, d)\{r, s\}>$, and a path connecting them through intermediate chain e. This path has connectivity 2, indicating that two moves are required before the terminal groups are safely connected, and is formally described as:

$$2\text{-}path(<(a, b)\{p, q\}> <(c, d)\{r, s\}> (e) \{p, q, r, s, t, u, v, w, x, y\})$$

Empty points $\underline{v}$ and $\underline{y}$ are underlined to indicate that they were pivot points during a step extension operation involved in creating this path. These are the two moves required to safely connect the terminal groups.

Paths are *commutative* and directionally insensitive, that is n-*path*($<C_1$ $\{S_1\}>$ $<C_2$ $\{S_2\}>$ C_3 $\{S_3\}$) and n-*path*($<C_2$ $\{S_2\}>$ $<C_1$ $\{S_1\}>$ C_3 $\{S_3\}$) are equivalent. Paths are created in either of two ways:

- *step extension spanning two disjoint groups, or*

- *combining existing paths using path operators.*

4.3.1 Paths from Steps

The first method of deriving a path is a special case of step extension. Instead of terminating on an empty pivot point, the step terminates on a point occupied by a chain belonging to a disjoint group, as follows:

$$n_1\text{-step}(<C_1 \{S_1\}> \underline{p} \{S_3\}) \quad \oplus \quad n_2\text{-step}(\underline{p} <C_2 \{S_2\}> \{S_4\})$$
$$\Rightarrow \quad n_3\text{-path}(<C_1 \{S_1\}> <C_2 \{S_2\}> \{S_5\})$$

$$iff \{S_1\}\wedge\{S_2\}\wedge\{S_3\}\wedge\{S_4\} = \{\}$$
$$and \quad C_1 \wedge C_2 = \{\}$$

$$where: \quad n_3 = n_1 + n_2 - 1$$
$$\{S_5\} = \{S_3\}\wedge\{S_4\}\wedge \underline{p}$$

The intermediate step point $\underline{p}$ is added to the empty point set of the resulting path, and its total connectivity is the sum of the component steps' connectivities minus one. This reflects the fact that the final step ends on an occupied point, so one less move is required to establish the connection.

Figure 4.6 illustrates the combination of 1-step($<a>$ $\underline{r}$ $\{p, q\}$) and 1-step($\underline{r}$ $<b>$ $\{s, t\}$) to give 1-path($<a>$ $<b>$ $\{p, q, \underline{r}, s, t\}$). This method of creating a path does not use any intermediate chains, so the intermediate chain set is initially empty and is not shown.

There may be several paths between two groups, but each such path must contain the minimal set of empty points to be valid. Paths with empty point supersets are superfluous and should be discarded.

4.3.2 Paths from Existing Paths

The other way in which paths may be created is through the combination of existing paths. Chains that do not belong to either terminal group may be used as stepping stones through which the path passes, forming intermediate chain sets.

As with steps, it is essential that empty point sets within paths do not overlap when paths are combined. Such an overlap would create vulnerable points that can be exploited by the opponent.

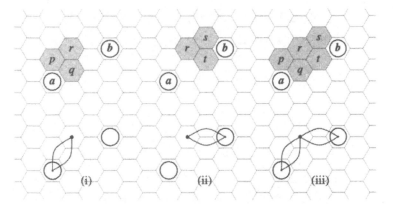

Figure 4.6. Two steps combine to form a path with pivot point r.

Figure 4.7 shows the 0-path(<*a*><*b*> {*p, q*}), the 0-path(<*a*><*c*> {*q, r*}) and a possible combination *n*-path(<*a*><*c*> {*p, q, r*}). The resulting path combination is vulnerable at point *q*, the overlap of the empty points sets, hence this combination is invalid and *n* is undefined. If the opponent plays at point *q*, they interfere with both paths with a single move.

Path operators analogous to the step operators previously described allow existing paths to be safely combined to create new paths.

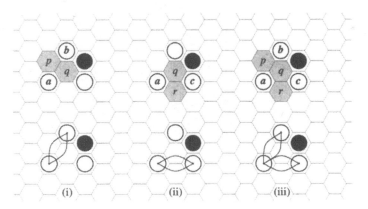

Figure 4.7. A vulnerable combination of paths.

4.3.3 Path Extension ⊕

This operator extends paths by concatenating a path from group <C_1 {S_1}> to group <C_2 {S_2}> with a path from <C_2 {S_2}> to <C_3 {S_3}>, if and only if all chain sets and empty point sets involved are disjoint.

n_1-path($<C_1\,\{S_1\}> <C_2\,\{S_2\}> C_4\,\{S_4\}$) $\oplus$ n_2-path($<C_2\,\{S_2\}> <C_3\,\{S_3\}> C_5\,\{S_5\}$)

$\Rightarrow$ n_3-path($<C_1\,\{S_1\}> <C_3\,\{S_3\}> C_6\,\{S_6\}$)

iff $\{S_1\}\wedge\{S_2\}\wedge\{S_3\}\wedge\{S_4\}\wedge\{S_5\} = \{\}$

and $C_1\wedge C_2\wedge C_3\wedge C_4\wedge C_5 = \{\}$

where: $n_3 = n_1 + n_2$

$$C_6 = C_2\vee C_4\vee C_5$$
$$\{S_6\} = \{S_2\}\vee\{S_4\}\vee\{S_5\}$$

Chains in the set C_2 belonging to the intermediate group $<C_2\,\{S_2\}>$ are added to the resulting path's intermediate chain set, and the empty point set $\{S_2\}$ belonging to the intermediate group $<C_2\,\{S_2\}>$ is added to the resulting path's empty point set.

The overall connectivity of the resulting path is the sum of the connectivities of the component paths. This allows a path to spread across the board, but weakens its connectivity in the process.

Figure 4.8 illustrates the extension of 0-path($<a> <b> \{p, q\}$) $\oplus$ 1-path($<b> <c> \{\underline{r}\}$) to give 1-path($<a> <c> (b)\ \{p, q, \underline{r}\}$). Empty point $\underline{r}$ is a pivot point, indicating that it is a key structural point within the path. This information is passed on to all future paths derived from this one.

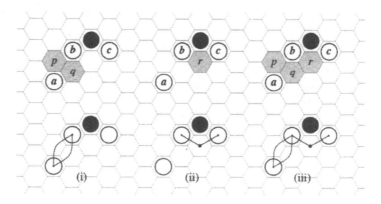

Figure 4.8. Path extension.

If the player faces the choice of a number of otherwise equally good moves, some of which are pivot points, then it is wise to play at a pivot point for a number of reasons:

- *it turns the component steps into paths, strengthening their connection, and*

- *it prevents the opponent from playing there and attacking a key point.*

4.3.4 Path Consolidation ⊗

Two paths of connectivity 1 that lie between the same pair of groups $<C_1\ \{S_1\}>$ and $<C_2$ $\{S_2\}>$ can be *consolidated* if their intermediate chain sets and empty point sets do not overlap each other or either path's terminal groups, to form a single path of connectivity 0. That is:

$$1\text{-path}(<C_1\ \{S_1\}> <C_2\ \{S_2\}> C_3\ \{S_3\})\quad \otimes\quad 1\text{-path}(<C_1\ \{S_1\}> <C_2\ \{S_2\}> C_4\ \{S_4\})$$

$$\Rightarrow\quad 0\text{-path}(<C_1\ \{S_1\}> <C_2\ \{S_2\}> C_5\ \{S_5\})$$

$$iff\ \ \{S_1\}\wedge\{S_2\}\wedge\{S_3\}\wedge\{S_4\} = \{\}$$

$$and\ C_1 \wedge C_2 \wedge C_3 \wedge C_4 = \{\}$$

$$where:\quad \{S_5\} = \{S_3\}\vee\{S_4\}$$

$$C_5 = C_3 \vee C_4$$

Figure 4.9 illustrates the consolidation of 1-path($<a> <b> \{p\}$) ⊗ 1-path($<a> <b> \{q\}$) to give 0-path($<a> <b> \{p, q\}$). The resulting path safely flows around the opponent's piece and cannot be broken.

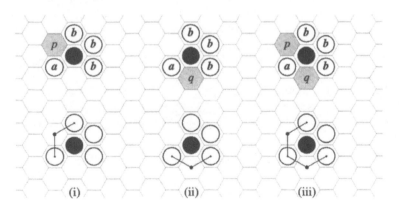

Figure 4.9. Path consolidation.

The order of connectivity of the resulting path is reduced by one, reflecting the fact that the player has an alternative route if one of the component paths is blocked. Path connectivity cannot be improved beyond 0-connectivity, hence it is futile to consolidate two 0-paths. This operator does not increase board coverage, but strengthens existing connections between groups.

It can be seen that each path consolidation operation creates a pair of *dual consolida-*

tion points formed by the single pivot points of the component 1-paths. If the opponent intrudes into either of these points then the path can be saved by playing its dual. The bridge pattern is the minimal path with dual consolidation points.

4.3.5 Spanning Paths

Any n-path($<e_0>$ $<e_1>$ C $\{S\}$) where $<e_0>$ and $<e_1>$ are the player's two edges is called a *spanning path* and gives a measure of the player's overall connectivity. If $n = 0$ then the player has effectively won the game with an unbeatable connection between their edges. It should be both players' aim to minimize their spanning path and to maximize their opponent's spanning wherever possible (see Basic Strategy, Section 3.1.4).

It's tempting to think that paths may be repeatedly consolidated without constraint, improving their connection, to yield a neat algorithmic solution for a given board position. However, this is not the case as a connection is only as strong as its weakest link. If threatened by an opponent, a connection of order 2 requires two replies for completion, a connection of order 3 requires three replies for completion and so on. But as each player only has *one move with which to reply before the opponent can again intrude*, such connections clearly have no guaranteed defense, hence only connections of order 1 may be consolidated safely.

This concept is illustrated in Figure 4.10. If unconstrained path consolidation were valid then the following would be true:

$$1\text{-path}(<a><c>\{p\}) \quad \oplus \quad 1\text{-path}(<c><b>\{r\}) \Rightarrow \quad 2\text{-path}(<a><b>(c)\{p, r\})$$

$$1\text{-path}(<a><d>\{q\}) \quad \oplus \quad 1\text{-path}(<d><b>\{s\}) \Rightarrow 2\text{-path}(<a><b>(d)\{q, s\})$$

$$2\text{-path}(<a><b>\{p, r\}) \otimes 2\text{-path}(<a><b>\{q, s\}) \Rightarrow 1\text{-path}(<a><b>(c, d)\{p, q, r, s\})$$

$$1\text{-path}(<a><e>\{t\}) \quad \oplus \quad 1\text{-path}(<e><b>\{v\}) \Rightarrow \quad 2\text{-path}(<a><b>(e)\{t, v\})$$

$$1\text{-path}(<a><f>\{u\}) \quad \oplus \quad 1\text{-path}(<f><b>\{w\}) \Rightarrow \quad 2\text{-path}(<a><b>(f)\{u, w\})$$

$$2\text{-path}(<a><b>\{t, v\}) \otimes 2\text{-path}(<a><b>\{u, w\}) \Rightarrow 1\text{-path}(<a><b>(e, f)\{t, u, v, w\})$$

$$1\text{-path}(<a><b>(c, d)\{p, q, r, s\}) \otimes 1\text{-path}(<a><b>(e, f)\{t, u, v, w\})$$

$$\Rightarrow 0\text{-path}(<a><b>(c, d, e, f)\{p, q, r, s, t, u, v, w\})$$

This implies that Black has a spanning path of order 0 and hence the win, which upon analysis is clearly not the case. This sequence is invalid because the 2-path components of

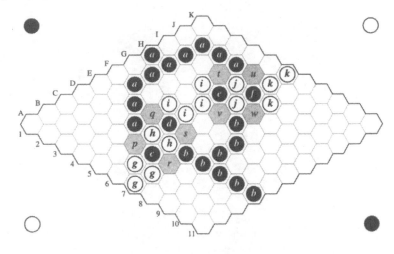

Figure 4.10. Unconstrained path consolidation is not safe.

1-path(*<a><b>*(*c, d*){*p, q, r, s*}) and 1-path(*<a><b>*(*e, f*){*t, u, v, w*}) are inherently weak and can not be consolidated safely.

In fact it is White that has a safe spanning path of order 0 and therefore the win. This leads to the observation that *for any given board position, a 0-connected spanning path may exist for either player but not both*. See Appendix D.5 for an informal proof of this.

If the player's spanning path is 0-connected then playing at pivot points within it will reduce the number of moves required to make the win more obvious. This may occur in an end game when the opponent has not realized that the game is lost.

4.4 Groups from Paths

Since all chain-to-chain connections within a 0-path must be 0-connected, including those between chains of the disjoint terminal groups and intermediate chain set, it follows that any path is equivalent to a safe group. A safe groups is derived from a 0-path as follows:

• *chains comprising the path's terminal groups and any intermediate chains used in the path's development comprise the safe group's chain set, and*

• *empty points within the path and its terminal groups combine to make the safe group's empty point set.*

Figure 4.11 shows a 0-path between the group *<(g, h, i)*{*t, u, v, w*}> previously introduced in Figure 4.1(iv) and group *<j>*. The path is comprised of the empty point set {*m, n, o̱, p, q, r, s, x̱, y, z*} and is the result of the consolidation of two 1-connected paths from the original group to *<j>*. The overall path between these two groups forms the new safe group *<(g, h, i, j)*{*m, n, o̱, p, q, r, s, t, u, v, w, x̱, y, z*}>.

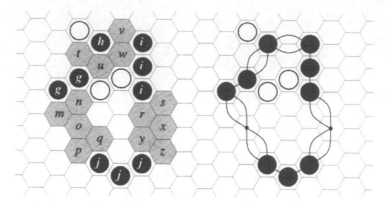

Figure 4.11. A 0-path between group *<(g, **h**, **i**){t, u, v, w}>* and group *<j>* forms 0-group *<(g, **h**, **i**, **j**){m, n, o, p, q, r, s, t, u, v, w, x, y, z}>*.

Safe groups can therefore be generated in either of two ways:

- *each chain (and edge) forms a safe singleton group, and*

- *each 0-path forms a safe group.*

This leads us back to the generation of new groups, and completes a cyclic mechanism for the iterative creation of connections across the board:

safe groups $\rightarrow$ steps $\rightarrow$ paths $\rightarrow$ 0-paths $\rightarrow$ safe groups $\rightarrow$ steps $\rightarrow$ paths etc.

This cycle is repeated until a spanning path is found and is the basis of the algorithm described in Chapter 8 that generates the best spanning paths for both players. Spanning paths give a direct measure of a player's connectivity across the board, how close they are to a win (relative to their opponent), and key structural points within the spanning path.

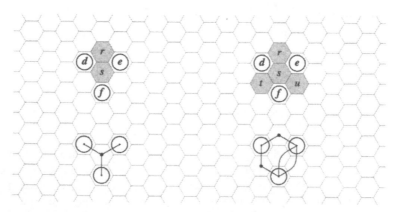

Figure 4.12. The empty point set is critical in describing a group's structure.

As a final point, it should be noted that empty point sets are critical in describing a group's structure. Using step extension and path consolidation, we can now show that the unsafe configuration of points shown in Figure 4.1(iii) becomes a safe group with the addition of two more empty points *t* and *u* (see Figure 4.12). This structure has trilateral symmetry and is equally valid whether the bridge connection occurs between piece pairs *d-e*, *e-f* or *f-d*.

Summary

Safely connected sets of chains form groups, from which steps and paths may be derived. Steps are essentially unrealized paths.

Steps provide an understanding of the potential network of connections across the board, and help pinpoint focal points that represent good moves. Given the choice between a step and the equivalent path, however, the path is obviously the stronger of the two as both terminals (its two groups) are realized (the pieces have been played).

Path generation can be performed iteratively to determine the best spanning paths for each player.

Templates

A connection template on the Hex board is a predefined pattern of pieces with a certain guaranteed connectivity. Templates dramatically reduce the complexity of board analysis by providing readily recognized situations that do not require further analysis.

5.1 Connection Templates

David Boll describes a *connection template* as "a pattern of open hexes that will allow connection even if the opponent has next move" [1994]. Templates are comprised of:

• *two terminal points (or a terminal point and a terminal edge) which may be occupied, and*

• *a set of empty points, which may contain pivot points.*

The bridge pattern introduced in Section 2.3.1 is an example of a template that occurs frequently during a game. Templates describe predefined board patterns with known connectivity, which can be exploited to speed up the path growth and board evaluation algorithms. As with groups, paths, and steps, points belonging to the template's empty point set *must* be unoccupied for the template to be valid.

There are two distinct types of template:

• *interior templates that describe a connection between two points, and*

• *edge templates that describe a connection between an edge and a point.*

5.1.1 Interior Templates

Interior templates represent a pattern of connectivity between two points. The *terminal points* between which the connection occurs may be either empty or occupied, provided that all pieces involved are of the same color. Unoccupied terminal points form *pivot points* from which further connections may be made. Template regions are shaded in the following examples.

Interior templates are comprised of adjacent and bridge moves from a point, as illustrated in Figure 5.1. Points *p* and *q* are the terminal or pivot points, and *s* and *t* along the bottom row are members of the bridge's empty point set. There are six rotations for each interior template.

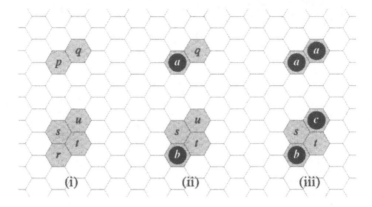

Figure 5.1. Step extension, step, and path templates for adjacent (top row) and bridge (bottom row) moves.

Column 5.1(i) shows an adjacent connection and a bridge connection between two empty points. These templates are used by the step extension operation (Section 4.2.2).

Column 5.1(ii) illustrates interior templates between a piece and an empty point, equivalent to the adjacent step *1-step(<a> q)* and bridge step *1-step(<b> u {s, t})*. These are the initial steps from groups before the step consolidation and extension operators are applied (Sections 4.2.1 and 4.2.2). Note the absence of any empty point set in the description *1-step(<a> q)*.

Column 5.1(iii) shows interior templates between two pieces, giving the chain *<a>* between adjacent pieces, and the bridge path *0-path(<b> <c> {s, t})* which is equivalent to the safe group *<(b, c){s, t}>*. Note that the hexagonal distance for the upper (adjacent) terminal pieces is one, while the hexagonal distance between the lower (bridge) pair of pieces is two, although they are both of equal connectivity 0. The bridge is clearly a superior way to increase connectivity with a single move.

Figure 5.2 shows the connection links for these templates explicitly. Adjacent points are linked *directly* as in the upper row of Figure 5.2. Bridge connections are linked *indirectly* via the safe path between them, as shown in the bottom row. The analysis of connection links within templates is useful for understanding more complex templates. In keeping with the convention used throughout this book, pivot or step points are dotted.

Templates may be devised for more complex combinations or for greater distance between terminal points, but it so happens that these more complex situations can be constructed iteratively from the simple templates shown.

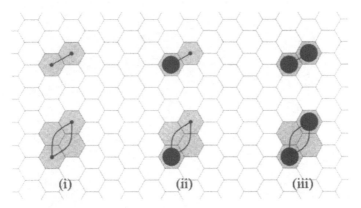

Figure 5.2. Interior templates with connection links shown.

5.1.2 Edge Templates

Edge templates are distinct in that one of their terminals is a board edge. They provide a ready means of recognizing how well a piece or point is connected to that edge.

Figure 5.3 shows an edge template for connection between a White piece and a White edge. Shaded hexagons indicate the template's total area, including empty points and terminals. As with interior templates the empty point set must be empty for the template to be valid.

The links within the template 5.3(i) are shown in Figure 5.3(ii), giving some insight into its structure. It can be seen that each pivot point (dotted) is 0-connected to the edge by a bridge, and that the piece is safely connected to both pivot points. Since there is no overlap in either branch of the path leading from the piece to the edge, there are no vulnerable points that the opponent can exploit.

No matter where the Black moves within the template shown in Figure 5.3 on their next move, White can reply to secure the connection. This template is a safely connected 0-path between the terminal piece and the edge. If the terminal point were unoccupied then this template would describe a 1-step from the edge to the point.

The ten forms of safe edge templates are illustrated in Figures 5.4 and 5.5. They are labeled by hexagonal distance from their terminal point to the edge: distance 1 for the

Figure 5.3. An edge template: (i) empty point set, and (ii) links and pivot points.

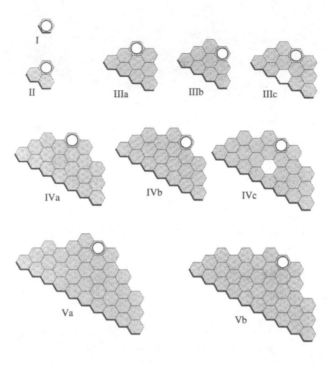

Figure 5.4. The ten minimal safe edge templates.

trivial edge template I, 2 for template II, 3 for templates IIIa, IIIb, and IIIc, 4 for templates IVa, IVb, and IVc, and 5 for templates Va and Vb. Templates IIIc and IVc each have a *don't care* point indicated by a white hexagon. This point may be empty or occupied by either playing without affecting the template's connectivity. There are two rotations for each edge template for both players, one corresponding to each home edge. Edge template pairs IIIa-IIIb, IVa-IVb, and Va-Vb are reflections of each other.

The edge templates shown are *minimal* in that they require the smallest possible empty point set for their safe connection. This is important during a game of Hex, where every point is crucial; the smaller the template, the better the chance of placing it effectively. A number of other safe templates that occupy larger areas exist, but these are supersets of the minimal templates and should not be used.

The underlying structure of the templates is revealed in Figure 5.5, which shows the links and pivot points within the template. As can be seen in all cases (except for IVa, IVb, Va and Vb) the template is safe as there are two or more non-overlapping subpaths branching from the piece to the edge. If Black intrudes in one subpath, White simply connects the other subpath.

The pivot points themselves do not provide enough information to allow the reconstruction of the template's internal subpath structure, but they do indicate weak points within the template. Pivot point information can be used to:

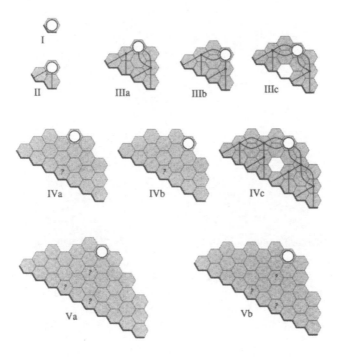

Figure 5.5. Links and pivot points within the ten safe edge templates. Points marked ? indicate intrusion points of interest.

- *attack opponent's templates at their most vulnerable points, and*

- *speed up the end game by filling in the player's own templates at the pivot points.*

5.1.2.1 The Asymmetrical 4-Row Edge Templates

The asymmetrical 4-row templates IVa and IVb require special explanation. Links and pivot points within these templates are not shown in Figure 5.5, as their safe connectivity is not immediately obvious. They are composed of overlapping subpaths that fortuitously combine to allow a safe connection, as shown in Figure 5.6.

The asymmetrical 4-row template IVa is shown at the top left of Figure 5.6 with a key overlap point marked ?. Figures 5.6(ii) and 5.6(iii) illustrate two subpaths between the edge and the terminal point that overlap at point ?. Figure 5.6(iv) shows a third subpath based on template IVc that flows around the overlap point ? should Black move there. In other words, the template is safely connected as *there are no points at which Black can play that intrude on every subpath*. This analysis also holds for the template's mirror reflection IVb.

This may appear to invalidate the condition described in Section 4.2.1.2 that paths may only be consolidated if their empty points sets do not overlap, and require the new

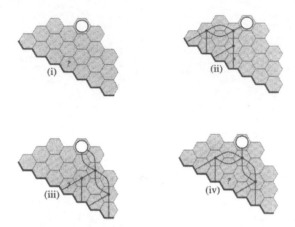

Figure 5.6. Three overlapping subpaths combine to make this template safely connected.

condition that *any set of paths may be consolidated if they do not all overlap at any point.* However, this rarely occurs in play except for edge templates and would drastically increase the complexity of path analysis, so the simpler rule stands.

The other edge template that warrants further discussion is the large 5-row template V.

5.1.2.2 The Asymmetrical 5-Row Edge Templates

Templates Va and Vb appear to be very strong patterns as they reaches five rows into the board (almost half the required distance for an 11x11 board). However they rarely occur during play on an 11x11 board, and are of limited use for a number of reasons:

•   *They require a large number of empty points (31 or over a quarter of an 11x11 board). Their large perimeter makes it vulnerable to adjacent plays and encroaching forcing moves.*

•   *They require almost an entire edge on the 11x11 board.*

•   *Their terminal points are poorly placed for further connection. The template area surrounds the terminal on each shoulder so play from this point can only occur in two directions.*

Pivot points are not marked for edge templates Va and Vb in Figure 5.5(ix) because proof of the templates' safety is nontrivial. Three key intrusion points are marked *?* in each case. If Black tries to block the template at any point other than these key points, White's defense is straightforward as shown in the following example. The following discussion focusses on template Va, but also holds for its reflective counterpart Vb.

The pattern on the left of Figure 5.7 shows a 5-row template and its three weakest points *p*, *q* and *r*. Other points within the template are trivial to defend, as any play by Black within these points can be answered by connecting a smaller subtemplate. This is shown in Figure 5.7(ii), where Black has intruded with move *1*, and White has replied with move *2*. Note that piece *2* is safely connected to the edge via template IVa, and that the

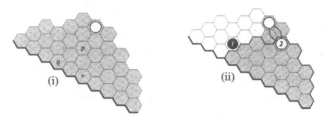

Figure 5.7. The large edge template Va and a safe reply to an intrusion.

overall 5-row template's terminal point is safely connected to *2* by a bridge. The subtemplate and bridge do not overlap, hence the terminal point remains safely connected to the edge and the template holds for intrusion *1*. Similarly, such immediate subtemplate replies can be found for all intrusions within the 5-row template except at points *p*, *q*, and *r* which require further analysis.

Figure 5.8(i) shows Black intruding into the template at point *p* with move *1*, and White's reply *2* to this intrusion. Black must now play within the region marked *b* to avoid White connecting directly to their terminal point.

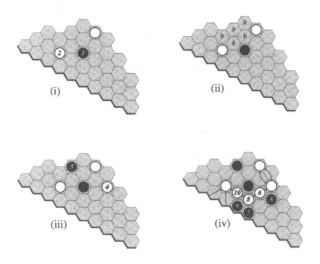

Figure 5.8. Template Va intrusion at point *p*.

Assume that Black chooses move *3*, as shown in Figure 5.8(iii). White replies with move *4* to any move *3* within this region. Figure 5.8(iv) shows the series of forcing moves that result in the terminal point's connection to the edge, which Black is unable to stop.

Similarly, Figure 5.9 shows White's defense to an intrusion into template Va at point *q*. White's best reply to the intrusion *1* is the bridge move *2* as shown in Figures 5.9(i) and

5.9(ii), to which Black has two practical replies. The sequence of forcing moves by White that defeats each of these moves *3* is shown.

If Black instead chooses to intrude into the template at point *r*, then White's best reply is move *2* as shown in Figure 5.9(iii). The two shaded hexagons indicate the region in which Black must move next turn to stop piece *2* from linking to the template's terminal piece and completing the connection.

If Black makes move *3* as illustrated in Figure 5.9(iv), White can complete the template easily with the sequence of forcing moves shown.

If Black plays *3* at the remaining vulnerable point as shown in Figure 5.9(v), then White's defense is more complicated. Black's move 5 is an attempt to disrupt both the connection of White's piece *2* to the edge, and the White subtemplate IIIb that threatens to

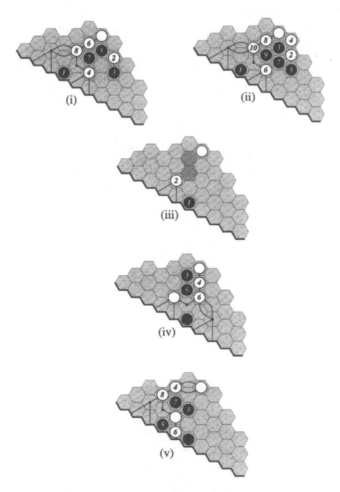

Figure 5.9. 5-row template intrusion at point *q*.

form on the left side. This move also attempts to gain extra board territory while intruding into a connection, but to no avail (see Section 6.3.1). White is again able to force the connection with the sequence of play shown.

This has exhausted Black's offensive possibilities; the template cannot be defeated as it stands (but it may be defeated if Black can contrive to intrude with a forcing move that requires White to play elsewhere on the board). Sequences of moves other than those described above may be attempted, but this will be of no avail as Figure 5.9 shows optimal play on Black's part.

Few Hex players are aware of templates Va and Vb, which can occasionally be used to trap the unwary. Perhaps their strategic importance will increase as they becomes more widely known and used.

5.2 Template Intrusions

A *template intrusion* occurs when the opponent moves within a template's empty point set. The template is then invalid unless the player replies with a move that restores the template's connectivity. Template intrusions are a key point of Hex strategy.

A *friendly intrusion* occurs when a player moves within their own template, in which case the template is also considered invalid. Its connectivity has not been threatened but the presence of an additional piece of the same color means that the template *can be described more succinctly by a set of smaller subtemplates*. This is useful for reducing complexity when constructing the connectivity graph across the board.

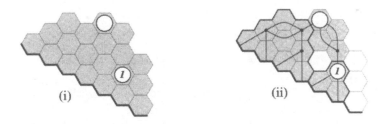

Figure 5.10. A friendly intrusion by White invalidates the original template, but is still safe.

Move *1* in Figure 5.10 is a friendly intrusion into White's edge template IVa. As can be seen from the link analysis in Figure 5.10(ii), this intrusion allows the terminal point to connect to the edge safely with a smaller empty point set (shaded) which constitutes a superior combination of subtemplates. Note that this particular combination is used for illustrative purposes only–even superior templates exist for this configuration of pieces!

It may seem unduly strict to make a template invalid following a friendly intrusion as its connection can only have improved, but it is important the connection with the *smallest possible empty point set* be used. Templates that form supersets of other templates without improving connectivity should be discarded to reduce unnecessary complexity.

5.3 Multi-Piece Edge Templates

A much richer variety of edge templates can be created if more than one player's piece is included in the connection. Some examples of safely connected multi-piece edge templates are shown in Figure 5.11. Each template is safely connected to the edge via terminal piece *t*.

Figure 5.11. Multi-piece edge connections with templates and terminal points shown.

The total set of possible multi-piece templates is too large to catalog here; however, a particularly useful 2-piece edge template worth mentioning is shown in Figure 5.12. This pattern is safely connected to the edge, occupies a relatively small region, and occurs frequently in play. It is most useful for squeezing ladder escapes into narrow gaps on the board.

Figure 5.12. A useful 2-piece edge template on the fourth row.

The fact that this template is safely connected is not immediately obvious. For instance, Figure 5.13 shows that the two simplest connections from the terminal pieces to the edge overlap at two points p and q. These are potentially vulnerable points that must be examined. If it can be demonstrated that the opponent cannot defeat the template by playing in either of these points, then the template is safe.

This argument also applies to the template's reflected counterpart.

Figure 5.13. Breakdown into component connections reveals vulnerable points of overlap p and q.

If Black intrudes at point *p* with move *1* as shown in Figure 5.14, then White's reply *2* threatens a direct connection. Black's forced reply *3* allows White to play *4* and secure the link.

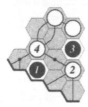

Figure 5.14. Solution to the intrusion at point p.

If Black instead intrudes at point *q* with move *1* as shown in Figure 5.15, White is able to follow a similar line of defense. Move *2* forces reply *3* from Black, setting up move *4* to complete the connection.

Figure 5.15. A solution to the intrusion at point q.

The possible number of combinations explodes with each extra piece so it is not practical to attempt to catalog all such templates here. However, a Hex-playing program may benefit by precomputing such templates up to a certain number of pieces, and storing this information for quick lookup.

Multi-piece edge templates are also useful in that the player can search for templates that are complete *except for one piece*. This missing piece directly describes a move that might be worth considering next turn. This approach is especially useful in identifying potential ladder escapes (see Section 7.3).

Summary

Templates specify board patterns that define guaranteed connections between points, pieces, or edges. Safely connected templates cannot be blocked unless the opponent intrudes with a move into the template, and the player is forced to play elsewhere by the intruding move.

Larger templates that require extra empty points may be derived from the set of defined safe templates, but these are superfluous and should be ignored. Only minimal templates need to be considered.

Laddered edge templates provide additional forms of safely connected edge templates. Considering multi-piece templates provides an even wider variety.

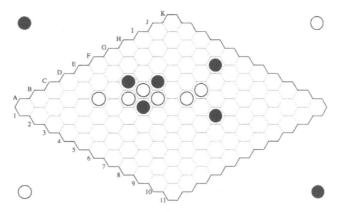

6

Strategy II: Intermediate

The following sections discuss points of strategy and tactical play that require some understanding of path generation (Chapter 4) and templates (Chapter 5). The illustrative board positions have been taken from actual games where appropriate, to give a feel for the context of that strategy within the game.

6.1 Expand With Templates

Templates form the basis of a player's connection across the board. A good understanding of templates allows a player to advance their connection quickly with optimal play.

Most games of Hex can be reduced to the connection of safely connected groups in the middle of the board to edge templates. Hex may therefore be described as the problem of:

- *establishing a connection to each edge, and*

- *joining these connections.*

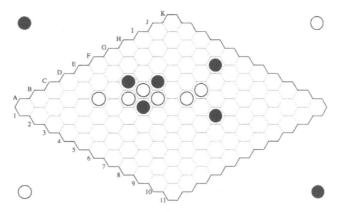

Figure 6.1. A typical game. Who is winning?

Many players follow this approach blindly, either first playing an edge template to either edge then trying to join them, or developing a strong force across the middle of the board that they then try to connect to each edge. A well-rounded player will maintain a more flexible approach, adopting these strategies to the situation at hand. A more holistic approach to the game that strives to reduce the player's weakest link in their best connection across the board is more successful.

For an example of template analysis, consider the game illustrated in Figure 6.1.The interior connection templates for White are shown in Figure 6.2. It can be seen that White's two most distant pieces, *x* and *y*, are safely connected by a series of adjacent and bridge templates.

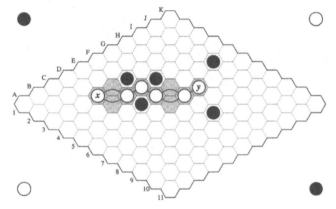

Figure 6.2. Interior templates.

Figure 6.3 shows that pieces *x* and *y* are also safely connected to their nearest home edges. Template IVb connects to the bottom left edge and template IVc connects to the top right edge. Recall that IVc is one of the two templates that contain an internal gap or *don't*

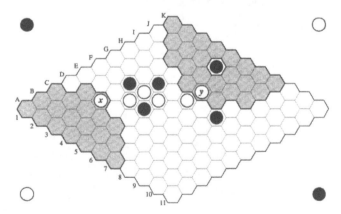

Figure 6.3. Edge templates.

care point, which makes it extremely useful on a cluttered board; the Black piece at J5 fits neatly into template IVc's *don't care* point.

The interior and edge templates can be joined at pieces *x* and *y* without overlap as shown below in Figure 6.4, and are hence safe. White now has a 0-connected spanning path giving them a winning connection.

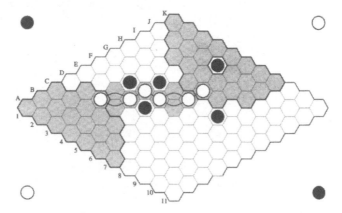

Figure 6.4. White has a winning path.

Black cannot hope to win this game and it would be diplomatic to resign shortly. It cannot hurt to play an extra move or two to make sure that White also realizes the win, but it is a waste of time to continue play until a solid chain spans the two edges. Unfortunately many games of Hex drag on long after the outcome is known due to misplaced tenacity on one player's part.

For smaller boards, edge templates that span the entire board and connect both edges are possible. Such edge-to-edge templates are not discussed here as they represent trivial solutions to the game for the player moving first; the game becomes nothing more than a

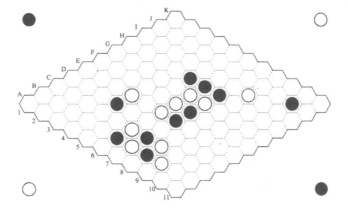

Figure 6.5. Board position to be analyzed.

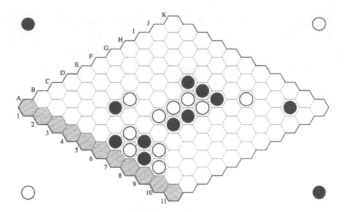

Figure 6.6. Points actually adjacent to White's lower left edge.

series of template intrusions and replies, until the opening player wins. Enforcing the swap option makes things a bit more interesting, but not enough to warrant further analysis of smaller boards.

Given the choice of a number of otherwise equally good edge templates, the player is generally wise to choose the template with the smallest empty point set.

6.1.1 Visualize Effective Edge Connections

It is possible to further clarify edge connections by visualizing the set of points adjacent to the edge to consist of both points actually adjacent to the edge and those points adjacent to chains safely connected to the edge. This technique will be demonstrated using the board position shown in Figure 6.5

First consider White's connection relative to their lower left edge. Points that are actually adjacent to this edge are indicated in Figure 6.6.

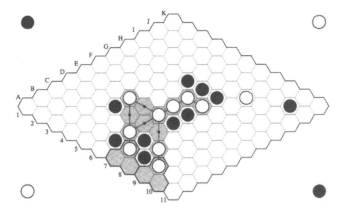

Figure 6.7. A White group safely connected to lower left edge.

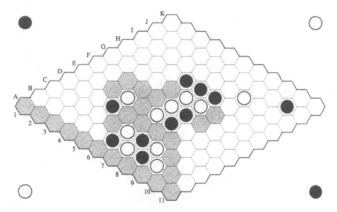

Figure 6.8. Points effectively adjacent to White's lower left edge.

Figure 6.7 shows a group of White chains safely connected to the edge. Links between chains and intrachain pieces within the group are shown. Although the group contains several pivot points (dotted) it is demonstrably safe as each pivot point has a dual.

Points *effectively* adjacent to White's lower left edge are shown in Figure 6.8. These points can be considered edge points in the sense that any White connection from the upper right edge that reaches any of these points is a winning spanning path.

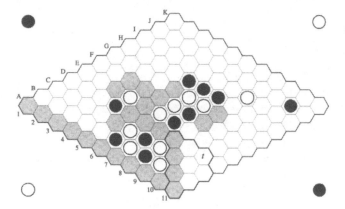

Figure 6.9. Template IIIa applied to the effective edge points.

Figure 6.9 shows how effective edge points can be used for edge template matching as safely as actual edge points can be used. Here we see edge template IIIa with terminal point *t* safely attached to the effective edge set, hence safely attached to the edge. This template does not overlap the safely connected group's empty point set at any point, otherwise this template would be invalid in this situation.

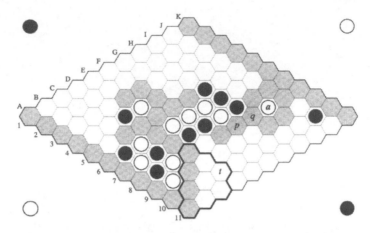

Figure 6.10. Points effectively adjacent to both of White's edges.

Figure 6.10 shows those points effectively connected to both of White's edges. Piece *a* is safely connected by all three 3-row edge templates, so points adjacent to it are marked as being effectively adjacent to the upper right edge. Points adjacent to the template's empty points but not adjacent to either the edge or the terminal piece *a* are not included.

Note that the two effectively adjacent point sets touch at points *p* and *q*, indicating that White is in a very strong position. Indeed, White is now in an unbeatable position as shown in Annotated Game 11.5.

Thanks to Tom Hayes for pointing out this useful technique.

6.2 Momentum

The player who is dictating play at any given point in the game is said to carry the *momentum* of play. Alternatively, the momentum is against the player who is forced to answer a series of attacking moves with no choice of reply.

Hex is a discrete game where momentum often swings between players with each move. This is most obvious towards the end of a closely-matched game, where White may be one move away from a win following their move, then Black one move from a win following their move and so on. This process may continue until either player makes a mistake or a killer move and the game ends. This constant pressure to force a winning/losing move within a constrained space makes Hex a very unforgiving game, causing one player to describe it as a "knife fight in a phone booth" [Boll 1997].

The amount of momentum that swings with each turn is generally more pronounced the smaller the board size, as each player is fewer moves from a win from the very start. This effect becomes less prevalent for boards of size 14x14 and larger, when the game becomes less discrete in nature.

Except when answering a forcing move, the player with the turn in hand usually has the advantage. This is another reason moving first is such an advantage: it gives the first player the initial momentum, as well as the fact that the piece may come in useful later in the game. Swapping the first move is a trade-off between losing the momentum (which goes to the opponent who may then play a threatening move in the center) and having a ladder escape that may save the game later.

Waiting to see what the opponent is up to before responding is not always a good idea, as the player hands the momentum of the game over to the opponent.

The game annotated in Appendix A.2 is a good example of a dramatic change in momentum that wins a game. Black loses the momentum from the start of the game and faces defeat during a long series of forcing moves. However, White stumbles at move *44*, and with move *45* Black is able to swing the momentum and set up a win.

6.3 Forcing Moves

Forcing moves are moves that immediately threaten a win or a strong connection, to which the opponent is forced to reply on their next turn. This includes template intrusions, to which it is usually wise for the opponent to respond immediately. The reply to a forcing move that offers little or no choice is called a *forced reply*. A winning series of forcing moves is shown in the following example:

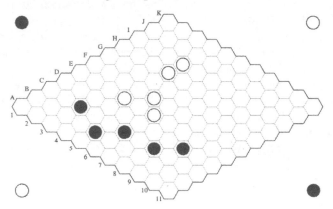

Figure 6.11. White to play and win.

The board position shown in Figure 6.11 does not look especially good for White at first glance. However, with their next move they are able to initiate a series of forcing moves that results in a win.

White plays *11* B6 to which Black must reply *12* C5 or face immediate defeat (Figure 6.12). Move *11* is an example of a forcing move, and move *12* is a forced reply. This play enables White to place a piece on the opposite side of Black's defensive line that is connected to bottom left the edge. Black, with almost no choice of reply, is not able to defuse this dangerous situation.

White then plays forcing move *13* C7 to which Black is forced to reply *14* A8 to again

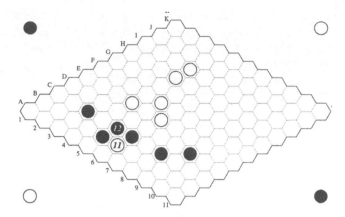

Figure 6.12. Forcing move *11* B6.

avoid immediate defeat (Figure 6.13). White then has an easy win with *15* B7, as revealed by the 0-connected spanning path shown in Figure 6.14.

This exchange demonstrates the importance of momentum within a game; in an otherwise close game, *the player with the move in hand has the advantage.* If it were Black's turn to play move *11* then they would have won with *11* C7.

Forcing moves are the only way to successfully intrude into an opponent's template. Templates with connectivity 0 are considered safe in isolation, but the following example demonstrates how they can be defeated by an intrusion that is also a *more threatening* forcing move.

White's edge template connecting the piece at D7 to the lower left edge in Figure 6.15 is safe in isolation, but consider what happens if Black moves at point p. White is forced to ignore the intrusion and move at point q to avoid immediate defeat. The template intrusion remains unanswered, and the template is no longer safe.

Note that this type of forcing move intrusion into a template does not work if the

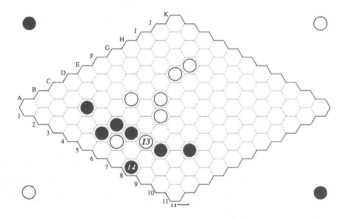

Figure 6.13. Forcing move *13* C7.

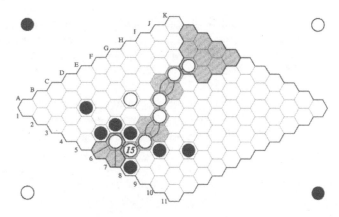

Figure 6.14. Move *15* B7 gives the victory.

template is part of a 0-connected spanning path. This is because the rules concerning path growth (discussed in Chapter 4) guarantee that the component parts making up a path do not overlap and can hence be defended successfully. A 0-connected spanning path cannot be beaten.

The following options are available to the player when the opponent has just made a forcing move, as suggested by David Boll [1994]:

- *answer the forcing move and save the link,*

- *give up the link and move elsewhere (if not a winning link), or*

- *play a forcing move themselves.*

The appropriate choice for a given situation depends largely on the severity of the forcing move. If it is just a nuisance move that threatens to break a minor link, it may be ignored.

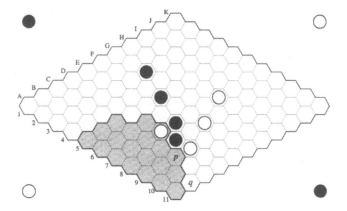

Figure 6.15. Forcing moves are the only way to successfully intrude into the opponent's template.

If the forcing move is a potential game-winner it must be answered immediately. In general *a forcing move should not be ignored unless the reply is a stronger forcing move itself or the threatened connection is not essential.* When considering a reply to a forcing move, the player should first determine how important the link is to their overall connection and whether it can be abandoned or not.

Forcing moves are a good way to gain the momentum, and when used properly force the opponent into a series of weak forced replies. This is a good opportunity for the player to force a win or develop their connection while maneuvering the opponent into a disadvantageous position.

It is worth noting that some players resent having the play dictated to them, and will choose a weaker reply rather than concede the expected forced reply. Such players can often be manipulated into losing positions through only a few such moves.

6.3.1 Stealing Territory

Recall from Section 2.1 that empty points adjacent to a piece are said to be *touching* that piece. The entire set of touching pieces for either player is described as that player's *territory*.

Figure 6.16(i) shows a board position with two Black pieces safely connected by a bridge across points p and q, and two adjacent White pieces. The central column shows White's territory *Tw* for this position (shaded), and the column on the right shows Black's territory *Tb* (shaded).

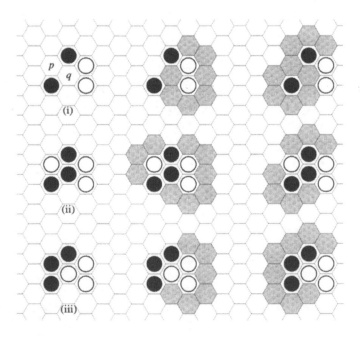

Figure 6.16. White and Black territory for three board positions.

For this example, *Tw*=7 and *Tb*=9. An estimation of White's *territorial strength*, indicating the amount of territory owned by White relative to the amount owned by Black, is given by:

$$Tw - Tb = -2$$

The board position shown in Figure 6.16(ii) results if White intrudes into the bridge at point p and Black is forced to reply at its dual point q to complete the connection. For this situation *Tw*=9 and *Tb*=7 and White's territorial strength is +2.

Figure 6.16(iii) shows the board position that results if White intrudes into the bridge at point q instead, and Black is forced to reply at its dual point p. For this situation *Tw*=6 and *Tb*=8 and White's territorial strength is -2.

The situation illustrated in Figure 6.16(ii) is clearly the best result for White: their territorial strength improves from -2 to +2, effectively stealing four touching positions through the use of an appropriate forcing move. In addition, the quality of their territory has improved, as they now touch points on both sides of Black's connection. Additional territory can never hurt the player, and increases their options for future threats.

With this play White has increased their territory without otherwise disturbing their connection or their opponent's. They have effectively stolen territory at no cost.

The definition of a player's territory can be modified to include not only empty points that touch a player's piece, but also empty points that are a bridge move away from a player's piece. This gives broader board coverage and is useful in that it indicates potential connections, but is a weaker definition.

6.3.2 Overlapping Bridges

A situation to be wary of is the case of overlapping bridges, as illustrated in Figure 6.17.

Suppose that it is Black's turn. Let's examine the consequences of their intruding into White's bridge by playing in both vulnerable points.

If Black intrudes into the non-overlapped point with move *1*, as shown in Figure 6.17(ii), then White's reply *2* in the overlap point both:

- *completes the White bridge, and*

- *intrudes into Black's bridge.*

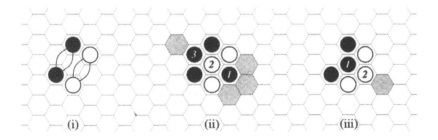

Figure 6.17. Overlapping bridges and two resolutions of play following Black's move 1.

Black is now forced to play *3* to restore their own bridge. Black has gained some additional territory (shaded), but also lost a move and given the momentum to White. In addition, the piece *1* is boxed in by White pieces and is now of limited use. This is not a good result for Black.

If Black instead intrudes into the overlapped bridge point with move *1* as shown in Figure 6.17(iii), then White is forced to reply with move *2* if they wish to restore their bridge. White's territory has increased by one point following this play (shaded), while Black has gained nothing.

From this example it can be seen that no advantage is to be gained by intruding into the opponent's overlapping bridge, when considered in isolation. It's generally best to leave this formation alone unless some outside influence such proximity to an edge, orientation, or the placement of nearby pieces allows a more worthwhile result.

6.4 Home Area

The board can be roughly divided into *home* and *enemy* areas relative to each player. The border between the two is defined by the topology of the board, and is the limit at which a player directly threatens to connect to their edge.

Figure 6.18 shows a sequence of bridge moves by Black that eventually connect to the bottom right edge via an edge template. White attempts to block unsuccessfully with a sequence of bridge moves. If this play had occurred one row away from the edge then Black would not have been connected; Black's home area should not extend beyond this line of play.

Figure 6.19 shows a better play by White. Instead of blindly following Black to the edge, White blocks the potential bridge with *4* H8. This is a ladder formation, and Black can force the ladder by moving at *p*. Although White has impeded the attack they have been forced to concede a 3-row ladder; this line is clearly a strong one for Black, and is a sensible choice for the limit of Black's territory. Ladders are discussed in detail in Chapter 7.

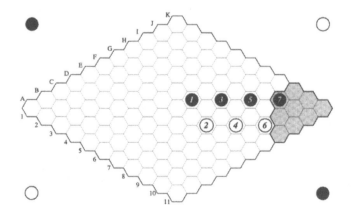

Figure 6.18. Black bridges down to the edge.

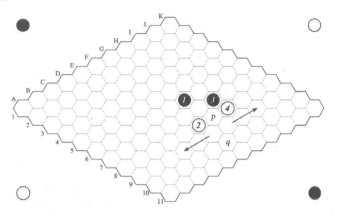

Figure 6.19. White blocks but concedes a ladder.

A division of board area that has proven useful is illustrated in Figure 6.20. Shaded hexagons represent Black's home area (border points are marked *b*) and white hexagons represent enemy areas. The key central hexagon is marked *c* and represents the bottleneck that joins the player's two home areas. The four corner hexagons marked *n* are neutral and are not considered to belong to either player's area.

Moving within the home area is generally more attacking; the player's edge is within reach and connections are easier to make. Moving within an enemy area is generally defensive, and most often occurs when a player is trying to block the opponent or responding to some immediate threat.

The home border points (marked *b*) are optimal positions that the player should strive to occupy if possible. They are points from which it is possible to both attack and defend at the same time. The vertical home border is the board's short diagonal, whose merits were briefly discussed in Section 3.2.1.

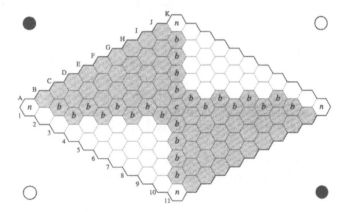

Figure 6.20. Board areas relative to Black: home (shaded), enemy (white), border (b), and neutral (n).

Notice that the player's home area is divided into two main regions corresponding to each of their edges, which join together through the bottleneck at F6. The player that controls this narrow passage has the distinct advantage as they are in a good position to connect to both edges. The player who does not control the central passage is forced to detour around it into enemy area in order to connect their two edges.

6.5 Edge Awareness

A piece occupying the central point F6 is 1-connected to both edges as shown in Figure 6.21. The required template is shaded gray and pivot points to connect to each edge are marked. There is only one combination of step and template that achieves a 1-connection to each side, which the opponent should be able to impede or block entirely if they play well.

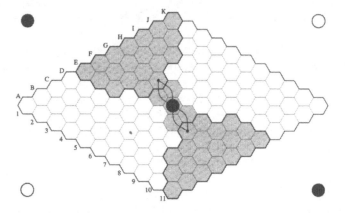

Figure 6.21. F6 is 1-connected to both edges. Template areas and pivot point shown.

Now consider a single piece played at C6 as shown in Figure 6.22. This piece is played along the same row as F6 but shifted towards the opponent's edge by three points, leading to a significantly different situation in terms of connections to each side. It is still 1-connected to the top left edge, this time by several combinations of a single step and an edge template. For clarity, only one such combination is shown. White will have difficulty blocking this connection.

However, the piece at C6 is at best 2-connected to the lower right edge. Several combinations of two steps and an edge template allow this, but all paths pass through the point p. If White were to play there next turn, Black would have difficulty in connecting C6 to the lower right edge.

Given that a player's overall connection is only as strong as the weakest link in their best path, Black's position in the first example is far superior to their position in the second example. Having two good connections is preferable to having an excellent connection and a poor one.

This example demonstrates that *straying towards an opponent's edge limits the opportunities available to the player*. Not only is the player's home edge further away, but

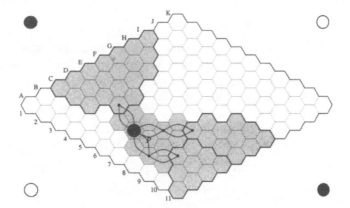

Figure 6.22. C6 is 1-connected to one edge, but 2-connected to the other.

the closer they stray towards the opponent's edge, the greater the opportunity that the opponent can block the way with an edge template. Playing across an opponent's template and breaking through it is possible (see Figure 6.15), but extremely difficult and requires considerable planning.

Another important aspect of edge awareness is observing how a point's proximity to the nearest edge reduces the number of adjacent points that it touches, as shown in Figure 6.23. Empty points adjacent to the pieces on the board are shaded. It can be seen that:

- *piece **a** occupies an interior point and has 6 adjacent neighbors,*

- *piece **b** occupies an edge point and has 4 adjacent neighbors,*

- *piece **c** occupies an obtuse corner and has 3 adjacent neighbors, and*

- *piece **d** occupies an acute corner and has 2 adjacent neighbors.*

The more adjacent points touched by a piece, the greater its connective potential and value in the game.

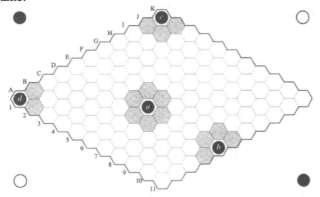

Figure 6.23. Proximity to an edge reduces the number of touching points.

6.5.1 Know Your Neighborhood

Recall from Section 2.1 that a point's immediate neighborhood consists of those adjacent points with which it shares an edge. Before moving in an empty point, it is worth looking at its relationship to its immediate neighbors. This can be quickly evaluated by traversing the empty hexagon's edges and adding 1 unit if any edge is touched and 1 unit for every chain and empty point that is encountered.

This *neighborhood value* gives an indication of the point's relative worth in the game. The neighborhood values for the various types of points (interior, touching one edge, obtuse corner, and acute corner) are shown in Figure 6.24.

Figure 6.24. Neighborhood values for interior, edge, obtuse corner, and acute corner points.

Various cases of neighborhood values for empty points touching at least one piece are shown in Figure 6.25. Schensted and Titus point out that empty points with a neighborhood value of three do not have any effect on the game and would constitute a wasted move, calling this formation the Worthless Triangle [1975]. This is irrespective of whether the surrounding pieces are the player's or the opponent's.

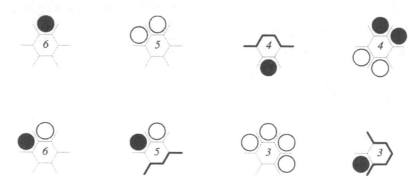

Figure 6.25. Neighborhood values for various empty points.

The more regions that touch a point, the higher its likelihood of being a useful play in general. Points with a neighborhood value of:

- *3 regions or less are worthless,*

- *4 regions are sometimes worthless and sometimes useful,*

- *5 regions are more often useful, and*

- *6 region are generally useful.*

An empty point's neighborhood value is useful for indicating danger points that may be given low priority. However, this is a very general strategic point and the player should evaluate their position carefully rather than simply rely on this value to indicate the worthiness of a move.

6.6 Loose Connections

Adjacent moves provide a guaranteed connection but cover little ground. Bridges provide connections over a greater distance and are *almost* as strong. What is the next best connection if even more distance is required?

To cover more distance than a bridge, a step from a pivot point is required. The best single step connection is composed of a bridge step and an adjacent step and is called a *loose connection*, as illustrated below. Two adjacent steps are arguably better connected, but can cover no more distance than a bridge.

Figure 6.26 shows two pieces removed by a bridge and an adjacent step. There are two combinations of steps that achieve this connection, the links of which are shown. Both combinations overlap at points p and q which are the vulnerable points in which the opponent must intrude to defeat this connection.

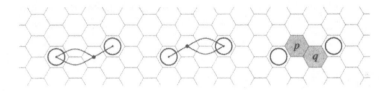

Figure 6.26. Bridge and adjacent steps form loose connections, with vulnerable points shown.

The loose connection threatens to connect the two pieces, puts pressure on the opponent and narrows down their options, and covers more ground than an adjacent or bridge move. It is also less tangible than a direct connection, and may cause the opponent some problems. Recall from Section 3.1.3 that it's best to defend at a distance. The *classic defensive* move described is actually an adjacent and bridge step away from the approaching piece, the same relationship as the loose connection.

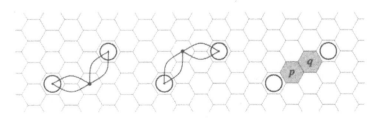

Figure 6.27. A pair of pieces connected by two bridge steps at an angle.

Even more ground can be covered if the two pieces are connected by a pair of bridge steps as illustrated in Figure 6.27. However this combination is easier for the opponent to beat. Its component steps overlap at points p and q and the whole pattern has a significantly larger empty point set than the bridge/adjacent combination, making it more vulnerable to intrusion from nearby plays. Figure 6.28 shows a connection between two pieces with an even greater separation composed of two bridge steps in line. As there is only one link it is particularly vulnerable at pivot point r.

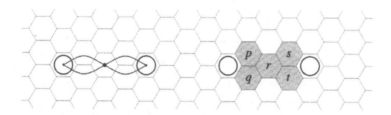

Figure 6.28. A pair of pieces connected by two bridge steps in line.

This section highlights the trade-off between connectivity and distance that players should be aware of as they expand their connections across the board. They should strive to achieve the optimum balance between these factors to achieve good board coverage without spreading their pieces too thinly.

6.7 Reduce the Opponent's Alternatives

Connectivity is based on the presence of alternatives; should one path be intruded into, an alternative path may be taken to maintain the connection. It's in the player's interest to reduce the number of alternatives available to the opponent as much as possible. This

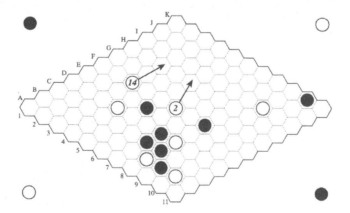

Figure 6.29. Two of White's forces threaten to combine into a strong assault.

applies equally to exact connections within a path and less well-defined connections that span large regions.

Figure 6.29 shows a board in which White finds themselves in a rather difficult situation, but threatening to combine attacks from *2* F6 and *14* F3 to develop a strong assault on the top right edge.

A good play for Black might be *15* G4 as shown in Figure 6.30. This move not only impedes the stronger attack from piece *14*, but also splits both assaults. White is forced to commit themselves to one attack or the other, rather than a more threatening concerted assault. Black has reduced the alternatives available to White by forcing them to commit to a specific line of play, impeding White's freedom to maneuver play to their advantage.

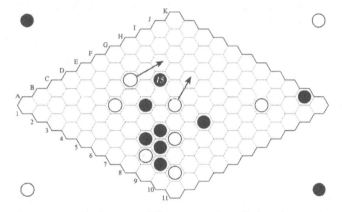

Figure 6.30. Black successfully counters by splitting the attacks.

6.8 Edge Defense

Following on from the previous example, it is now imperative that White stops Black piece *15* from connecting to the edge, as shown in Figure 6.31. Black wins the game if they make

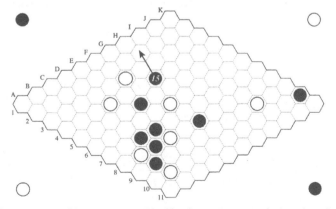

Figure 6.31. White must stop Black's piece *15* connecting to the edge.

this connection, as piece *15* is safely connected to their body of pieces in the lower part of the board, which is in turn safely connected to the bottom right edge by two ladder escapes. See the Section 7.7.2, Cascading Escapes, for an explanation of how these two ladder escapes combine to give a safe connection.

White's best play is to identify potential edge templates available to Black and to locate vulnerable points of overlap within them.

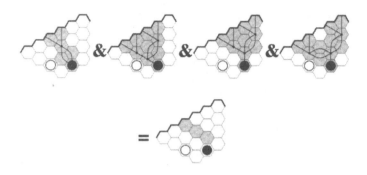

Figure 6.32. Black's edge templates overlap at three points.

Figure 6.32 shows a slice taken from the relevant area of the board shown in Figure 6.31. The top row illustrates those edge templates that are one step away from Black's piece *15*. These represent the winning moves available to Black next turn.

The bottom row of Figure 6.32 shows the three points at which all of the above templates intersect (shaded). If White plays outside this group then Black will win next turn. White is forced to play at H1, H2, or H3 on their next move.

Moves H1 and H3 will most likely result in a Black ladder along row 2, while move H2 will result in a White ladder down column J. White must decide which of these alternatives is in their best interests.

6.9 Stages of Play

It would be convenient to be able to divide a given game of Hex into distinct stages that can be individually analyzed. However, the complexity of the game and other factors such as discrepancies between the standard and playing style of the opponents conspire to make this task difficult.

Most Hex games quickly develop a unique character of their own as they progress. Opponents of similar skill and familiar with each other's playing style are often surprised at how quickly two games will diverge from almost identical starting conditions. It is useful to have some awareness of the stages that may occur in a game of Hex, and the best strategies for those stages.

This section outlines stages that tended to unfold most commonly in a broad survey of Hex games. Percentages of moves made are rough guides only; the actual percentage is not known until the last move is made. However, these measurements give some indication of the relative length of each stage.

Disclaimer: These stages apply to a theoretical game between two conservative players of average skill; actual games seldom follow this pattern exactly.

6.9.1 Opening Moves

Moves: First three or four moves (including swap)

State of Play: The opening player should play a move that promises to be useful later in the game, but not so threatening that the opponent decides to swap. Section 9.1.3 discusses this further and describes under what circumstances the second player should swap.

6.9.2 Early Game

Moves: Up to 25% of moves made.

State of Play: Players try to gain as much territory as possible with sparsely spread pieces that are weakly connected. The aim here is to achieve an even connection across the board, while blocking the opponent's potential connections. These are the formative moves that shape the rest of the game, and are generally defensive in nature.

6.9.3 Middle Game

Moves: 25% to 66% of moves.

State of Play: Potential spanning paths are more firmly established and each player tries to strengthen their connection and weaken the opponent's. This stage is generally defensive with the occasional aggressive move to create openings. Some forcing moves and preparations for ladder escapes occur here.

6.9.4 End Game

Moves: 66% to 90% of moves.

State of Play: By now potential spanning paths are well defined and each player tries to complete/break connections with close-in combinatorial play. As board coverage increases, empty points that do not belong to either player's territory become rare. Hence this stage involves mostly forced moves and ladder formations, and is aggressive in nature. Momentum can swing dramatically with each move at this point, and the first fork or other advantageous move will usually win the game.

6.9.5 Closing Moves

Moves: Remaining moves.

State of Play: The game has been decided with one player completing a 0-connected
                      spanning path. The losing player may make a few redundant moves within
                      the winning connection to make sure that their opponent is aware of the
                      win. Experienced players will concede when this stage is reached.

Summary

Players should learn to identify connection templates within a given board position, and
use these to develop their connection. Edge connections can be visualized for clarity.

Players should strive to carry the momentum of a game, which can be achieved through
the use of forcing moves. Forcing moves can be used to steal territory at no cost to the
player and maneuver the opponent into a disadvantageous or losing position.

Players should be aware of the trade-off between connectivity and distance when expanding their connection across the board.

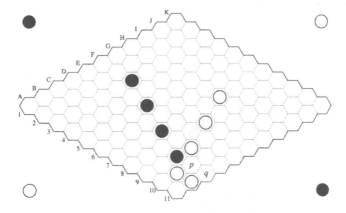

7

Ladders

Ladders and ladder escapes are a point of strategy that warrant a chapter of their own. Due to the cut/short nature of Hex, players will sooner or later find themselves hitting a defensive wall set up by their opponent. A good understanding of how to force ladders and escape from them will assist the player in getting around such impasses.

7.1 Ladder Basics

A *ladder* occurs when one player strives to force a connection to an edge but is deflected by the opponent a constant distance away, resulting in a sequence of moves in a direction parallel to the edge.

Figure 7.1 shows a ladder about to form. Black is forced to move at point *p* to avoid imminent defeat. This itself becomes a forcing move, and White is forced to reply at point *q*.

Figure 7.1. A ladder about to form along the bottom right edge...

If Black continues to press towards the lower left edge, White is forced to reply along the edge with each move, leading to the situation shown in Figure 7.2. Here Black has blindly played the ladder through to White's advantage.

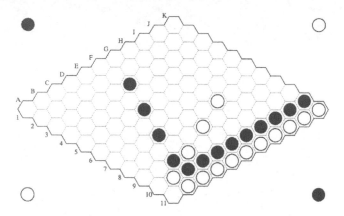

Figure 7.2. ... leading to a win for White.

The *row* of a ladder refers to the distance that the attacker is held away from the edge; hence, the ladder shown above is a 2-row ladder. 2-Row ladders are the closest possible ladders to the target edge and are generally the most threatening. Ladders may occur along any row up to half the board size; however, ladders further out than row 4 have a decreasingly small chance of being beneficial to the attacker.

The defender does not have to play in a straight line for ladders along row 3 or greater, and may deviate towards the edge, resulting in an $(n-1)$-row ladder for each move towards the edge. Figure 7.3 shows White defending what is initially a 4-row ladder, but after two bridge moves towards the target edge becomes a 2-row ladder. In general this is not a good idea, and it is best to *keep the attacker as far from the target edge as possible*. The

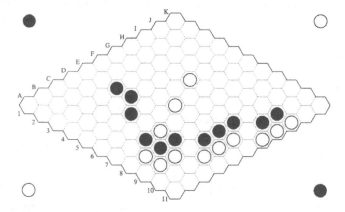

Figure 7.3. White allows a ladder to deviate towards the target edge.

defender cannot force the ladder *away* from the target edge, the best they can do is maintain a steady distance.

It can be useful to view a ladder as a single unit [Boll 1994]. Although the defender may deviate towards the edge for ladders along row 3 or greater, there will be only one or two optimal moves per turn along the ladder. It pays both the attacker and defender to play any ladder through mentally before committing to it; if the ladder favors the opponent, additional action is required.

7.2 Ladder Formation

Potential ladders can be identified before their actual formation by recognizing common patterns. Figure 7.4 illustrates one such pattern called a *bottleneck* that occurs often during play.

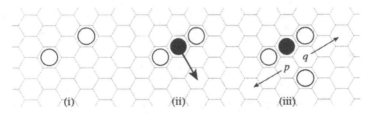

Figure 7.4. The common bottleneck ladder formation.

A bottleneck composed of two White pieces is shown in Figure 7.4(i). Black pushed through this bottleneck in the direction shown in 7.4.(ii), and in 7.4(iii) White plugs the gap. The light arrows indicate the two directions along which the ladder may grow from points p and q. The bottleneck formation has already been introduced in the discussion of template V in Section 5.3.2, Figure 5.8.

Figure 7.5 illustrates the bottleneck ladder pattern, and a couple of closely related bottleneck-bridge patterns. The starting points and directions of potential ladders resulting from these formations are shown. The bottleneck-bridge patterns are similar to the bottleneck pattern with the plugging piece removed by one position.

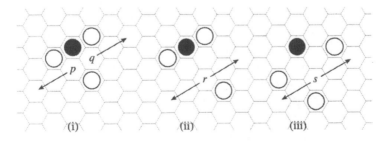

Figure 7.5. Bottleneck and bottleneck-bridge ladder formations.

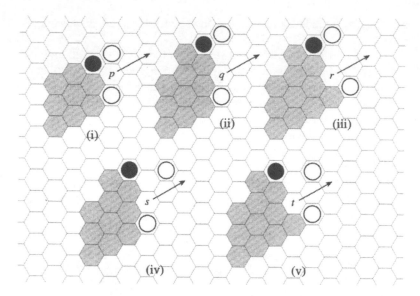

Figure 7.6. More general ladder formation patterns.

Bottleneck ladder formations are based on the more general patterns illustrated in Figure 7.6. Figure 7.6(i) shows the adjacent version, and 7.6(ii) shows the bridge version of these general patterns. In both cases the starting points and directions of potential ladders are shown, and shaded regions indicate regions within which there must be no attacker's piece that provides an immediate connection to the edge. If such a connection existed it would bypass the ladder, making it unnecessary. The ladder will only form if it is the attacker's best attempt at connecting to the edge.

7.3 Ladder Escapes

Consider what happens in the situation shown in Figure 7.1 if Black has one additional piece on the board at position J10, as shown in Figure 7.7. Again Black must move at point *p*, and White must reply with point *q*, forming a ladder.

The additional piece at J10 lies along the ladder, forming a *ladder escape* for Black. This allows Black to jump a move ahead of the ladder and complete their connection, as shown in Figure 7.8, winning the game. J10 is described as the *escape piece*.

The ladder escape is one of the most important Hex concepts, and is where the game is usually won or lost. Ladders and ladder escapes are the most common way to connect piece groups in the middle to each edge. A player's skill at Hex is usually determined by their ability to engineer successful ladder escapes.

To be successful, a ladder escape should:

• *be safely connected to the target edge, and*

• *not interfere with the ladder's projected path.*

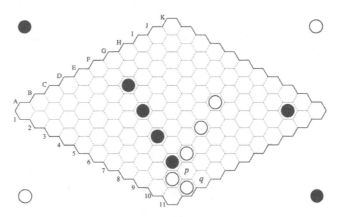

Figure 7.7. A ladder forms along the bottom right edge...

The danger of ladder escapes is another reason that the defender should strive to keep the ladder as far from the target edge as possible: the further out the ladder is, the smaller the chance of a successful escape.

Not every extra piece adjacent to the ladder's projected path is useful in assisting a ladder escape. Consider the board situation shown in Figure 7.9. White has forced a ladder from piece *a*. Piece *b* does not assist the ladder, however promising it may appear at first glance. This is because *b* is blocked in and touches only two empty points, *p* and *q*. By the time the ladder reaches point *p* to connect with *b*, it is already touching point *q* itself, hence there is no advantage from piece *b*.

A piece may be hemmed in by board edge or adjacent enemy pieces, reducing the number of empty points that it touches, such that it is no longer of any use to the player and is effectively removed from the game. Such a piece is described as an *embedded* piece. Schensted and Titus would describe this piece as lying within a Worthless Triangle (see

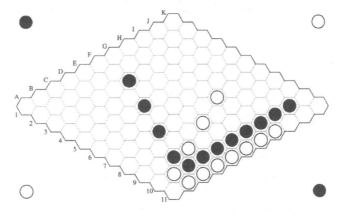

Figure 7.8. ... but Black escapes and wins the game.

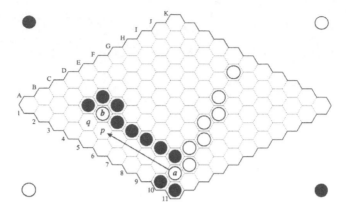

Figure 7.9. The embedded piece b does not assist the ladder.

Section 6.5.1). The fact that *b* is of limited use is suggested by its low neighborhood value of 3 (see Section 6.5.1).

In order to determine how close a potential ladder is to connecting to the edge, it is useful to recognize common ladder escape patterns. The following sections illustrate such patterns for the first three ladder rows, beyond which ladder escapes become less straightforward and do not have simple escape templates. Many other ladder escape patterns exist; those illustrated are the simplest templates with minimal empty point sets that have proven to be most useful.

7.3.1 2-Row Ladder Escape Templates

Figure 7.10 illustrates the minimal ladder escape templates for a 2-row ladder. The ladder formation is shown at the bottom of each Figure, originating at Black's ladder piece *l*, progressing along the direction indicated, and escaping at escape piece *e*. It's assumed that all points along the ladder's path are empty.

Observe that the escape templates shown closely match the edge templates listed in Section 5.3. The 5-row escape template is included here even though it is not possible to fit both it and a ladder formation on the same edge of an 11x11 board. This template does allow ladder escapes on larger boards under some circumstances, so should not be entirely discounted.

Note that the last five ladder escape templates in Figure 7.10 have some degree of overlap between the edge template's empty point set and the projected ladder path. For these specific cases, this overlap does not interfere with the ladder escape, but, in general, overlap between the edge template and projected ladder path should be avoided.

7.3.2 3-Row Ladder Escape Templates

Figure 7.11 illustrates the minimal ladder escape templates for a 3-row ladder. Only two of the nine original edge templates are suitable in this case. Points marked *x* must not contain any Black pieces or the ladder can be defeated, but may contain White pieces without affecting the ladder's outcome.

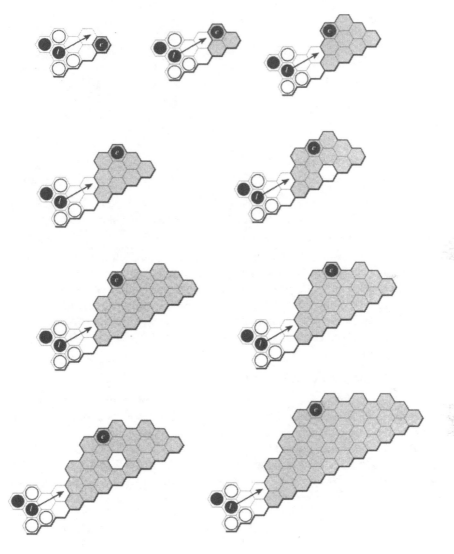

Figure 7.10. Ladder escape templates for 2-row ladders.

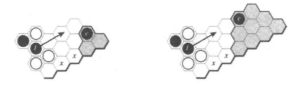

Figure 7.11. Ladder escape templates for 3-row ladders.

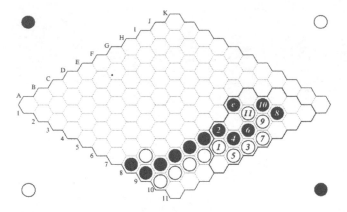

Figure 7.12. White can defeat this potential ladder escape template.

Figure 7.12 demonstrates that edge template IVa, the most likely 4-row candidate as it has the least amount of overlap with the approaching ladder's projected path, does not allow a safe escape. The sequence of play that defeats the ladder is shown.

The key play is White's *3* F11, which draws Black away from the potential escape piece *e*. This is one of the few situations where it's beneficial for the attacker to shift the ladder closer to the edge. The template is not shaded to indicate that it's suffered intrusion and is no longer safe.

7.3.3 4-Row Ladder Escape Template

Figure 7.13 illustrates the single minimal ladder escape template for 4-row ladders. Again, points marked *x* must not contain any Black pieces or the ladder can be defeated, but may contain White pieces without affecting the ladder's outcome.

Figure 7.14 illustrates how additional ladder escape templates can be derived from edge templates. A bridge move is made slightly off the ladder path that allows the attacker to connect with the escape piece. This simple concept allows the derivation of a large number of additional ladder escape templates, too numerous to list here.

Ladder escape templates should allow a player to quickly spot whether they have an obvious ladder escape. Often the situation is not so straightforward, and the player must use some analysis to determine whether the escape is viable or not. Notice that any ladder escape template that fails for a ladder on row *n* will also fail for ladders on rows *n+1* and greater.

Figure 7.13. Ladder escape template for 4-row ladders.

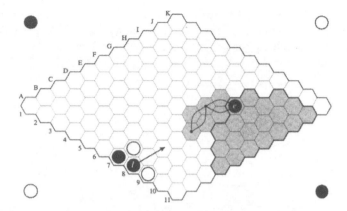

Figure 7.14. Additional 4-row ladder escape template derived by bridging to edge template IVa.

7.4 Ladder Escape Forks

If a player is forced into a ladder and no convenient escape is present, one must be engineered; blindly following a ladder to its conclusion is a losing play. The best way to create an escape is to play a forcing move that:

- *the opponent must respond to, and*

- *is safely connected to the target edge and lies along the ladder's path.*

This is an example of a *fork*, a concept that will be very familiar to players of Chess. A fork in Hex is a play that threatens to make two or more connections that do not overlap and hence cannot both be defeated by a single reply. A *ladder escape fork* is a fork in which at least one of the potential connections is a ladder escape. Schensted and Titus [1975] stress the importance of making each move serve at least two purposes, and call this the concept of Double Trouble.

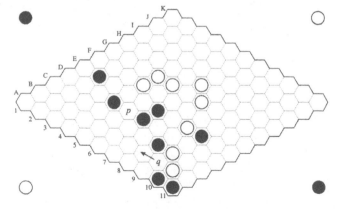

Figure 7.15. White to play.

Figure 7.15 illustrates a ladder escape in action. Black has a strong connection across the board with only two vulnerable connections at points p and q. Note that White is able to force a ladder from point q, but as yet no escape piece is present.

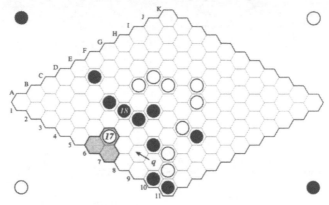

Figure 7.16. White plays a ladder escape fork with 17 B6.

White's move *17* B6 (Figure 7.16) is a ladder escape fork: it provides a ladder escape to the lower left edge via template II, and at the same time threatens to connect through point p to White's strong central connection at F4. These two threats do not overlap at any point.

Black is forced to play *18* D5 to avoid immediate defeat. White is now free to play at point q to form a 2-row ladder that will escape at piece *17* and complete their connection. *17* B6 has won the game for White.

Forks occur regularly throughout a game of Hex (the bridge pattern is a trivial case of a fork), but are at their most dangerous in the case of ladder escape forks. A properly played ladder escape fork guarantees the player a connection to that edge, and more often than not will win them the game. It's a killer move for which players should always be on the lookout and preparing for.

7.5 Ladder Escape Foils

In order to defend successfully against a ladder escape, the defender must either:

- *intrude on the ladder escape template, or*

- *block the projected ladder path.*

In the case of defending against a ladder escape fork, the situation is more dire and the defender must achieve both of these aims with a single move. This is demonstrated in the following example:

Figure 7.17 shows the same board situation as that illustrated in Figure 7.15, but in this case White has chosen to attempt a different ladder escape fork with *17* D3. This at first appears to be a good play as it threatens to connect with White's central group through point r, but is in fact vulnerable.

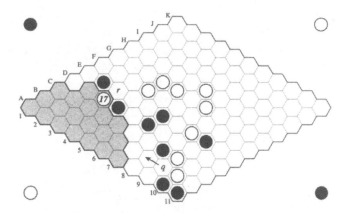

Figure 7.17. White plays a vulnerable ladder escape fork.

Figure 7.18 shows the sequence of moves that allows Black to defend successfully against the escape. *18* C4 is the critical move and is described as the *ladder escape foil*. It intrudes into both the ladder escape template and the ladder's projected path. White's reply *19* C2 reconnects the escape and maintains the threat, but Black is able to shut out both threats completely with moves *20* B3 and *22* C3.

White played poorly but can still win the game by playing the original ladder escape fork at B6. In general, *the smaller the ladder escape template and the less opportunity for intrusion, the stronger it is.*

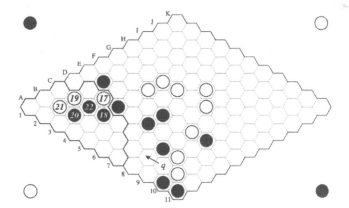

Figure 7.18. Black defends successfully against this potential escape.

Forcing moves, in particular forcing moves that intrude into enemy bridges, are the safest way to set up a ladder escape. It's therefore unfortunate that such forcing moves (for instance *17* D3) require that the potential escape piece be played adjacent to an opponent's piece and can hence usually be foiled.

7.5.1 Adjacent Move Foils

When exactly can a ladder escape be foiled, and when is it safe? Consider the board situation shown in Figure 7.19. White has forced a ladder along column B in the direction indicated, and has a potential escape in piece *y*. However, Black has a piece *x* adjacent to the escape piece *y* that lies in the direction from which the ladder originated. This piece *x* is sufficient to ensure that the ladder is foiled, as shown in Figure 7.20.

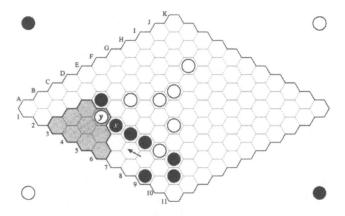

Figure 7.19. Situation in which a ladder escape can be foiled.

Black plays the ladder escape foil *1* B5, and White responds by restoring their connection to the edge with *2* B3. Black is forced to play *3* D4 but still blocks White's ladder and wins the game. The key to the success of move *1* is that it *intrudes into the potential escape's template with an adjacent move*. This leaves no vulnerable points that White may exploit as forcing moves to rekindle the threat.

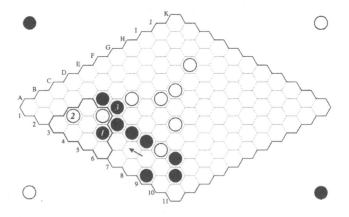

Figure 7.20. Black foils the escape.

Figure 7.21 shows a slightly different situation. In this case the closest Black piece to escape piece *y* along the ladder path, *x*, is separated by a single point *p*, which dramatically changes the outcome.

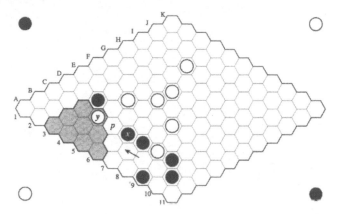

Figure 7.21. Situation in which a ladder escape cannot be foiled.

Figure 7.22 shows the sequence of play that results when Black attempts the same ladder escape foil *1* B5 that proved successful previously. This time it is a bridge move that leaves vulnerable points at C5 and B6. White replies with *2* B3 and Black is forced to reply with blocking move *3* D4 to avoid immediate defeat. White can then force the ladder with moves *4* B8 and *6* B7, then apply the *coup de grace* with *8* B6, which intrudes into the bridge from *x* and creates a winning fork between points *p* and *q*.

These examples illustrate that *a foiling move that leaves vulnerable points is likely to fail.*

To be successful, a foil must be played *before* the ladder reaches its escape piece, and in fact *must be played before the ladder is formed.* The temptation is therefore to play the

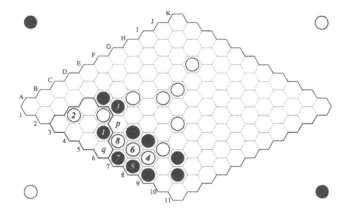

Figure 7.22. White can play through to force a winning fork between points p and q.

foil as soon as the escape piece is played. However, the player may choose to delay the foiling move until necessary. This is good gamesmanship in that the player forcing the ladder may not see the foil, and play erroneously elsewhere on the board under the false assumption that their ladder provides a guaranteed connection.

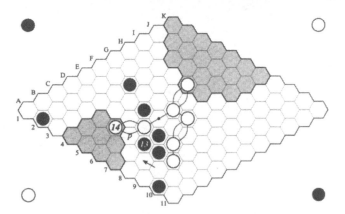

Figure 7.23. White looks to be in a winning position, but can be foiled using vulnerable point p.

7.5.2 Non-Adjacent Foils

It's possible to foil a potential ladder escape with a piece that is non-adjacent to the escape piece under some circumstances. For instance, consider the board position illustrated in Figure 7.23. White's ladder along column B appears to have a safe escape at piece *14*, but notice point *p* that is both adjacent to the escape piece *14* and lying on the forking path to White's main body of pieces.

Black may attempt to play directly in this vulnerable point with move *15* C6 as illustrated in Figure 7.24, but it is too late; White's ladder has already formed. Black is unable

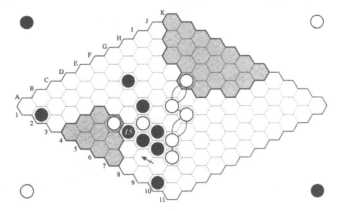

Figure 7.24. Bad play by Black gives White the win.

to intrude into the escape template with an adjacent move foil before White can push the ladder to its escape.

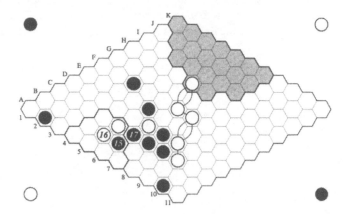

Figure 7.25. Black successfully foils the escape.

Figure 7.25 shows how Black should play to foil this escape. Move *15* B6 forces White to play a move such as *16* B5 to keep the escape alive, but Black is now able to play the forcing move *17* C6 that both intrudes into the escape piece's forking path and blocks the ladder, completing the foil.

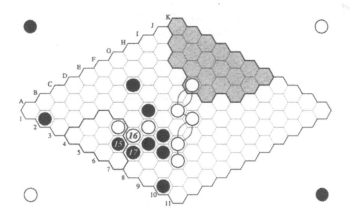

Figure 7.26. Black's foil is again successful.

Figure 7.26 shows a variation on the above play. If White attempts to avoid Black's forcing move by playing at vulnerable point *16* C6 themselves, then Black completes the foil with *17* B7.

7.5.3 Foiling From the Wrong Side

Figure 7.27 shows a situation similar to that shown in Figure 7.17. Piece *y* presents a potential escape for the ladder indicated, but there is an adjacent defender's piece *x* touching it in line with the ladder. Can the defender foil the ladder from this position?

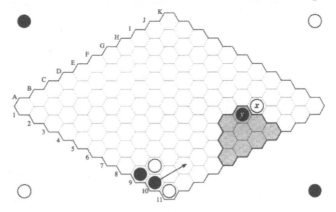

Figure 7.27. White attempts to foil from an adjacent piece on the wrong side of the escape piece y.

Figure 7.28 shows the result of White's defense from this position. Move *1* H10, the only adjacent play that intrudes into the ladder's template, can be refuted by a number of moves including *2* G10 which reestablishes the escape.

Black is now in an even a stronger position; the escape remains unfoiled, some territory has been gained, and there is now a larger chain from which to develop. White has gained territory but in a dead area near the opponent's edge. This is a very weak formation on White's part.

This example illustrates the concept that *ladder escapes cannot be foiled by pieces that lie beyond the escape template.*

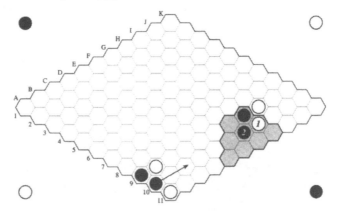

Figure 7.28. Black's escape is now stronger.

7.6 Getting Off Ladders

If playing through a ladder with no existing escape piece, an escape must be engineered as soon as possible. Figure 7.29 shows a typical situation. White has forced a ladder with moves *10* to *16*, but realizes that the ladder is doomed unless action is taken. *18* C5 is a valiant attempt at providing an escape: it threatens to connect to piece *4* via point *p*, though this is not really a fork given that Black could choose to ignore this mild threat and cut the ladder off at point *q*.

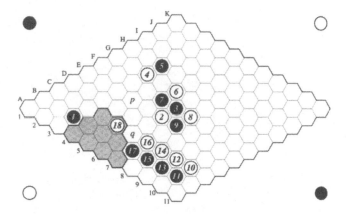

Figure 7.29. White engineers a ladder escape.

The best way to improvise an escape is to anticipate the ladder and play the escape piece well before it's needed, preferably in a different part of the board. If the move is relatively inconsequential and appears to be part of another strategy, it has a good chance of going undetected by the opponent. This is illustrated in Figure 7.30.

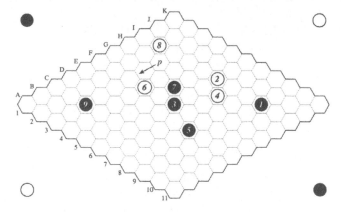

Figure 7.30. Black tries a potential ladder escape...

Black anticipates a ladder forming at point *p* and progressing in the direction shown in Figure 7.30, and plays *9* C3 to prepare an escape. This move occurs well away from the action and appears reasonably inconsequential at this point, but does present a minor threat of connecting to Black's pieces *3* and *5*. Will White fall for this trap and try to block the obvious connection, or will they try to foil the ladder escape?

White does indeed fall for the trap and plays the mediocre *10* D5 (Figure 7.31). They'd have been well advised to block the escape, perhaps with a move at H3. However, Black can now force the ladder with move *11* H3 and their connection to the top left edge is guaranteed.

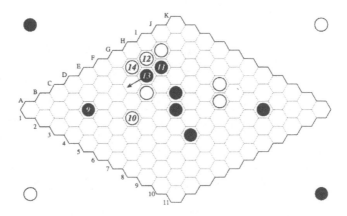

Figure 7.31. ... which White fails to counter.

If several choices of ladder escapes present themselves, the one with the most danger-ous (well-connected) forcing move should be chosen, as the defender will be more likely to answer the forcing move and leave the ladder escape intact. If there's still a choice, the escape with the smallest template area should be taken. This gives the defender less chance to intrude into the escape and foil it.

Forcing moves that steal territory often come in useful later in the game as ladder escapes, as shown in Figure 7.32. Forcing move *8* J5 gains territory for White, at the same time providing escape for any 2-row ladder that may later form near point *q*.

It may appear optimistic to view move *8* J5 as a possible escape with no evidence yet of a ladder forming; however, this move certainly does not harm White and could become quite important. It's generally a good idea to prepare as many avenues of escape as early as possible.

7.7 Partial Ladder Escapes

Some ladder escapes do not provide a direct connection to the edge, but are still useful as they allow the attacker to maneuver the ladder closer to the edge. Such plays are called *partial ladder escapes*, and can be used to devastating effect if the player is able to force a connection using the second (closer) ladder. This is illustrated in the following example.

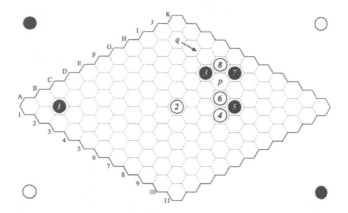

Figure 7.32. A forcing move that steals territory and prepares a possible ladder escape.

Figure 7.33 shows a board situation in which Black has forced at least a 4-row ladder with *11* E5. White plugs the gap with *12* F3, restricting it to a 4-row ladder. *13* H3 appears to be a feeble attempt at a ladder escape on Black's behalf (it does not threaten any additional connections) but is shortly revealed to be a dangerous play.

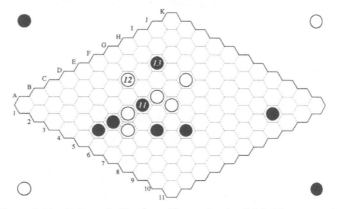

Figure 7.33. A 4-row ladder formation and potential ladder escape 13.

7.7.1 Foldback Escapes

White's move *14* H2 impedes the escape but, being a bridge, leaves potential forcing moves. Black reestablishes their escape piece connection with *15* I2. White's move *16* F4 appears to block the ladder escape once and for all... or does it? The arrow in Figure 7.34 indicates the path that Black will pursue in an attempt to maneuver around blocking piece *12* at F3 and connect with escape piece *13*, now securely connected to the edge. Black will use a partial ladder escape to achieve this[1].

---

[1] Note that White can block Black's partial ladder escape if they play *16* G3 rather than the more obvious *16* F4. This example illustrates a trap set by Black rather than optimal play by White.

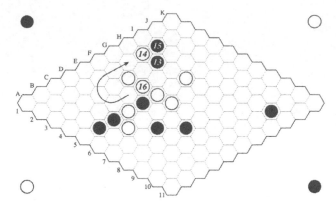

Figure 7.34. White foils then blocks the ladder escape, but Black can still use it in a roundabout
fashion.

Play proceeds as follows:

Black   White

17 E4   Black starts the 4-row ladder.

         18 E3   Forced reply.

19 D4   Black continues the ladder.

         20 C3   Again White is forced to reply here. D2 may appear to be a better defen-
                   sive play for White, but further analysis reveals it to be a losing move.

21 D3   Black pushes through White's defense to set up a 2-row ladder. The
                   partial ladder escape has been successful, and Black is free to pursue the
                   1-row ladder in the direction shown back towards escape piece *13*.

         22 E1   Forced reply.

     Black forces the ladder with moves *23* E2 and *25* F2, and wins the game with *27* G2
which forks winning moves at points *p* and *q* (Figure 7.36).
     The key point of this example is that an escape that fails for one ladder may still be
sound for another (closer) ladder. The fact that a partial ladder escape usually allows a
ladder to double back on itself is extremely useful, as the opponent may fail to pay sufficient attention to blocking potential ladder escapes on what appears to be the ladder's
weaker side. Such escapes are called *foldback* escapes.

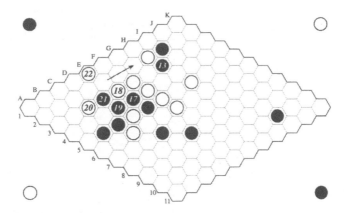

Figure 7.35. Black forces the partial ladder escape, resulting in a 2-row ladder.

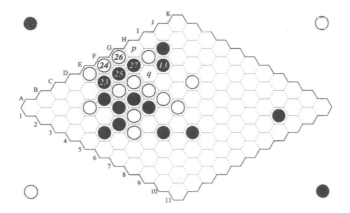

Figure 7.36. Black completes the 2-row ladder resulting in a win.

7.7.2 Cascading Escapes

A variation on the above example shows that partial ladder escapes do not necessarily double back on themselves, but can continue in the same direction as they force the ladder closer and closer to the target edge. Such partial escapes are said to *cascade* towards the edge, as demonstrated in Figure 7.37.

Recall from Figure 7.12 that the escape template e_1 can be defeated for 3-row ladders, and does not by itself provide an escape for the approaching ladder shown in Figure 7.37. However, White is forced to concede a 2-row ladder in their defense against this template as shown in Figure 7.38.

Black can capitalize on this situation by continuing to force the 2-row ladder, guaranteeing their escape via the secondary escape piece e_2 (Figure 7.39).

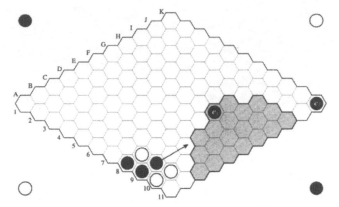

Figure 7.37. Escape template e_1 fails for 3-row ladders, but e_2 provides an escape.

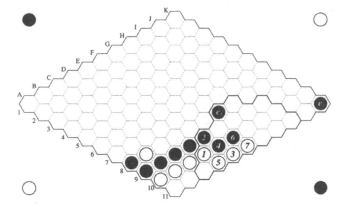

Figure 7.38. The ladder is forced to row 2 by template e_1...

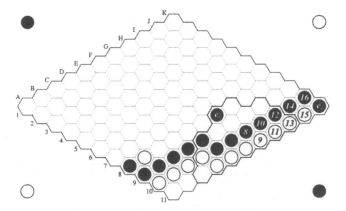

Figure 7.39. ... and escapes via escape piece e_2.

Summary

Ladders are one of the most important and involved aspects of Hex strategy. Players should spot the warning signs that point to an impending ladder formation, and plan escapes for potential where possible. The ladder escape fork is a killer move that will often win the game.

Ladder escape foils that leave vulnerable (non-adjacent) points are likely except under certain circumstances. Ladder escapes cannot be foiled from the side of the escape template opposite to that from which the ladder originated.

Seemingly inconsequential side plays may be turned into successful ladder escapes through the use of partial ladder escapes. Partial escapes may double back on themselves or cascade closer to the target edge.

Algorithmic Board Evaluation

Now that aspects of path generation, edge templates, and ladder escapes have been discussed in detail, it's possible to outline an algorithm for the accurate measurement of connectivity across the board for both players. An example board situation is then analyzed for both players to illustrate the algorithm in action.

8.1 The Algorithm

From any board situation the relative strength of each player's position is given by the strength of their best spanning path across the board. Chapter 4, Groups, Steps, and Paths, introduced the concepts behind spanning paths but did not explicitly state the method for deriving them. The following sections outline an algorithm that gives a precise measure of each player's connectivity.

The algorithm was implemented and tested in C++ and shown to be sound. It is presented below in a more universal pseudocode. Some knowledge of procedural programming languages is helpful in understanding the following sections but not essential. The algorithm is broken down and discussed in its constituent steps.

8.1.1 Notation

The pseudocode follows the following conventions:

- `Monospaced type` = pseudocode
- **`Uppercase()`** = subroutine or function
- **`lowercase`** = keyword
- *`lowercase`* = variable name
- *`var → set`* = add variable *var* to the collection *set*

An expression of the form **Sub**(p) indicates that subroutine **Sub()** has been invoked with parameter p. This is simply a way of parceling up work into smaller units that are easier to manage. **Sub**(p) may itself pass p onto further subroutines.

8.1.2 Main Approach

The main body of the algorithm determines chains of pieces belonging to each player, sets these as the initial groups along with the player's board edges, then progressively takes steps from each group until a spanning path is found. These iteratively-taken steps give rise to paths, which give rise to further groups.

Group collections and path collections such as *new_groups* and *new_paths* are *sets* in the programming sense of the word. They may contain a number of elements but no duplicates. Both sets are initially empty.

```
for each player p
    GenerateChains(p)
    new_groups = 0
    SingletonGroups(p, new_groups)
    new_paths = 0
    do
            TakeAStep(p, new_groups, new_paths)
    while not IsSpanningPath(p)
```

Two minor subroutines need only brief explanation:

GenerateChains(p)
        groups adjacent pieces into chains

The generation of chains is trivial. A new chain is created for each piece on the board that does not belong to an existing chain. All pieces that can be reached by adjacent steps from this piece are added to the chain.

IsSpanningPath(p)
        determines whether a spanning path exists for p

This subroutine checks the collection of paths to determine if any path has the player's two edges as terminal groups. Such a path is a spanning path.

The algorithm is guaranteed to find all minimal spanning paths. These may be ordered by the following criteria:

• *number of pivot points,*

• *number of empty points, or*

• *some arbitrary ordering, such as first found.*

Hence it's always possible to produce a best spanning path for either player. The algorithm may be set to terminate when either the first minimal spanning path is determined or all minimal spanning paths are determined.

8.1.3 Initial Singleton Groups

Recall from Section 4.1 that a singleton group is composed of a single chain. This definition is expanded to include single edges. Chains that touch an edge are deemed to belong to the edge and added to the edge's singleton group rather than form groups themselves.

```
SingletonGroups(p, new_groups)
    for each edge e
        singleton group from e → new_groups
    for each chain c
        if c touches an edge e
            add c to existing edge group
        else
            singleton group from c → new_groups
```

Following this operation all pieces and edges belonging to player *p* have been assigned to singleton groups, which have been added to the *new_groups* set.

8.1.4 Step from All Groups

A collective step is taken from each of the player's groups. This will usually result in new paths being created.

```
TakeAStep(p, groups, new_paths)
    new_steps = 0
    do
        for each new group g
            StepFromGroup(p, g, new_steps, new_paths)
            MatchPivots(new_steps, new_paths)
        CombinePaths(new_paths)
            IdentifyNewGroups(new_paths, new_groups)
    while new groups exist
```

New paths are combined with existing paths, possibly resulting in new groups being formed. New groups are simply detected as 0-connected paths.

```
IdentifyNewGroups(new_paths, new_groups)
    for each new path p
        if p is a 0-path
            group from path p → new_groups
```

8.1.5 Step from a Specific Group

Stepping from a particular group may require special actions depending on how many steps have already been taken from that group.

```
StepFromGroup(p, g, new_steps, new_paths)
    if first step from g
```

```
        TakeInitialStep(p, g, new_steps, new_paths)
else
        ExtendSteps(g)
if second step from g
        ConsolidateSteps(new_steps)
PathsFromSteps(new_steps, new_paths)
```

If this is the first step from the group, a special function **TakeInitialStep** is invoked, otherwise all existing steps from the group are extended. If this is the second step from the group, then step consolidation according to the rules outlined in Section 4.2.1 may occur.

New steps may or may not result in new paths.

8.1.6 First Steps from a Group

The first steps from a group require careful care. Special handling must be applied to groups that include an edge.

```
        TakeInitialSteps(p, g, new_steps)
        for each piece p in g
                adjacent steps → new_steps
                bridge steps → new_steps
        if g contains an edge e
                adjacent steps → new_steps
                bridge steps → new_steps
                templates to empty points → new_steps
             SafeTemplates(e, p) → new_paths
             SafeLadders(e, p) → new_paths
```

Firstly, steps are taken from all pieces within the group. This includes all adjacent and bridge steps available from each piece (see Section 4.2).

If the group also contains an edge, then additional steps to edge templates whose terminal points are empty are also taken. Edge templates whose terminal points are occupied by p are added directly to the *new_paths* set.

```
    SafeTemplates(e, p)
        finds chains safely connected to e by edge templates
```

Chains that form ladders safely connected to edge e are added directly to the *new_paths* set as well.

```
    SafeLadders(e, p)
        finds potential ladders safely connected to e
```

Determining whether a chain is safely connected to the edge by a ladder is difficult. It must first be determined whether the chain can in fact form a ladder, then whether the ladder can force an escape or not. There are two general approaches to achieving this non-trivial problem:

- *play out the ladder from each candidate piece, or*

- *reverse engineer possible ladders from the edge to reveal candidate pieces.*

8.1.7 Step Combination

Steps can be extended by making adjacent and bridge moves from their terminal points (Section 4.2.2). The previous terminal points then become pivot points within the new steps.

```
ExtendSteps(new_steps)
     for each new step s
          extend adjacent steps from s terminal →
          new_steps
          extend bridge steps from s terminal →
          new_steps
```

A pair of 2-steps that share the same terminal and do not overlap are consolidated to form a new 1-step. The new 1-step must then be extended to remain in synch with the group's remaining steps. This may result in recursive consolidation and extension of additional 2-steps.

```
ConsolidateSteps(new_steps)
     do
          for each pair of new 2-steps s₁ and s₂
               if SameTerminal(s₁, s₂) and
               IsDisjoint(s₁, s₂)
                    consolidate to new 1-step → new_steps
                    extend to new 2-steps → new_steps
     while new 2-steps exist
```

The terminal point of a step may be either:

- *a point,*

- *a chain, or*

- *an edge.*

```
SameTerminal(s₁, s₂)
     steps s₁, s₂ share the same terminal point/chain/edge
```

It is necessary to know whether pairs of steps overlap or are disjoint. They are disjoint if their empty point sets do not overlap at any point. The terminal point is not included in this calculation.

> **IsDisjoint(**step, step**)**
>   steps do not overlap, but may share common terminal

It is also necessary to apply similar tests of disjointness to chain-group pairs and pairs of paths. These items are disjoint if their empty point sets and constituent chain sets do not overlap.

> **IsDisjoint(**chain, group**)**
>   chain does not belong to group
> **IsDisjoint(**path, path**)**
>   paths do not overlap

8.1.8 Paths from Steps

The newly generated set of steps is then processed to determine whether any new paths are formed. Steps from group g that terminate on chains not belonging to g form new paths.

> **PathsFromSteps(**g, new_steps, new_paths**)**
>   **for** each new step s_1
>   **if** s_1 lands on c and **IsDisjoint(**c, g**)**
>   create new n-path → new_paths
>   **if** s_1 is a 2-step
>   **for** each new 2-step s_2
>   **if** s_2 lands on c and **IsDisjoint(**s_1, s_2**)**
>   consolidate to 0-path → new_paths

Any path created from a 2-step is a 1-path. Pairs of 2-steps from the group g to the chain c that do not overlap are equivalent to pairs of disjoint 1-paths that can be consolidated to 0-paths (see Section 4.3.4). Newly created paths are added to the *new_paths* set.

8.1.9 Path Combination

Conditions for combining paths by extension or consolidation are outlined in Section 4.3.

> **CombinePaths(**new_paths**)**
>   **for** each new pair of paths p_1 and p_2
>   **if CanExtend(**p_1, p_2**)**
>   extend to n-path → new_paths
>   **if CanConsolidate(**p_1, p_2**)**
>   consolidate to 0-path → new_paths

Two *n*-paths can be extended if they share exactly one common terminal group and are disjoint. The resulting path's connectivity is equal to the sum of the connectivities of *path$_1$* and *path$_2$*.

> **CanExtend(**path$_1$, path$_2$**)**
>   path$_1$ and path$_2$ share exactly one terminal group
>   **IsDisjoint(**path$_1$, path$_2$**)**

Two 1-paths can be consolidated if they share both terminal group and are disjoint. The resulting path is a 0-path.

> **CanConsolidate**(*path₁, path₂*)
>   *path₁* and *path₂* are both 1-paths
>   *path₁* and *path₂* both share both terminal groups
>   **IsDisjoint**(*path₁, path₂*)

8.1.10 Matching Pivot Points

Steps that originate from disjoint groups and converge at a common empty terminal point are combined to form a path, as illustrated in Figure 4.6. This is a type of step extension.

> **MatchPivots**(*new_steps, new_paths*)
>   **for** each new pair of steps *s₁* and *s₂*
>   **if** **SameTerminal**(*s₁, s₂*) and **DifferentSources**(*s₁, s₂*)
>   and **IsDisjoint**(*s₁, s₂*)
>   extend to n-path → *new_paths*

The resulting path's connectivity is equal to sum of the connectivities of the component steps.

> **DifferentSources**(*step, step*)
>   steps are from different groups that do not overlap

Two groups are deemed to be different sources if they do not share any common chains and their empty point sets are disjoint.

8.2 An Example

The algorithm described above will now be applied step-by-step to the board situation illustrated in Figure 8.1 as way of demonstration. Path evaluation is performed for both Black and White and key points are revealed by a comparison between the two.

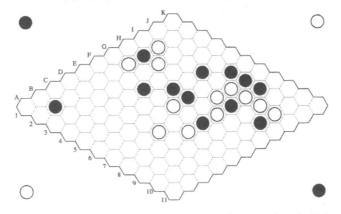

Figure 8.1. Sample game to demonstrate path generation algorithm.

To keep this example as clear as possible only those steps and paths that contribute to the best spanning path are shown. In reality, a much larger number of false steps and paths must be taken in order to determine the best path.

8.2.1 Path Evaluation for Black

The first stage in the evaluation of Black's position is to determine chains of Black pieces.

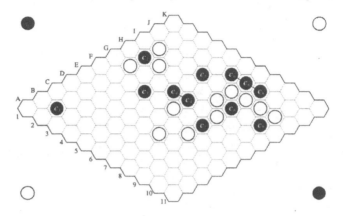

Figure 8.2. Black's chains.

Figure 8.2 shows Black's nine chains labeled c_1 to c_9. Links between adjacent pieces within chains are shown. These chains and the two Black edges form the initial set of singleton groups.

Initial steps are then taken from all groups. Figure 8.3 shows a template step taken from the top left edge. The step's terminal or pivot point is marked (dotted).

Figure 8.4 shows two edge template steps whose terminal points are occupied by

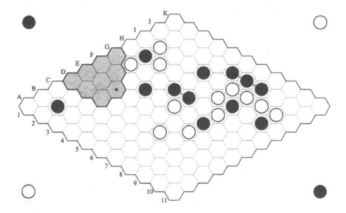

Figure 8.3. A template step taken from Black's edge with terminal (pivot) point marked.

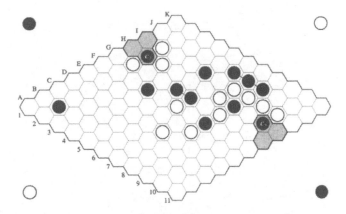

Figure 8.4. Chains safely connected by edge templates.

Black pieces. These form 0-paths (hence safe groups) between each piece and the edge. These groups do not overlap and can therefore coexist on the board.

A safely connected ladder formation from chain c_7 to the bottom right edge is shown in Figure 8.5. This is not formed by one of the ladder escape templates listed in Section 7.3, but forms a partial escape that can then take advantage of chain c_9 to complete the connection. The shaded area shows the minimum region required for this escape. This also forms a 0-path (hence safe group) between chain c_7 and the edge.

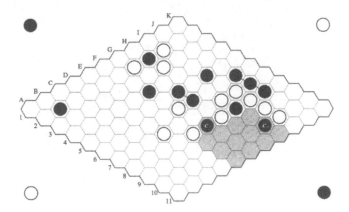

Figure 8.5. Safe ladder formation from chain c_7 to the edge.

Figure 8.6 shows initial steps taken from the singleton group formed by chain c_3. These include an adjacent step to G3, a bridge step to E3, and a bridge step to chain c_4. Pivot points are shown (dotted).

The bridge step to E3 meets edge template step shown in Figure 8.3 and the two steps are combined to create a 1-path from chain c_3 to the edge. The adjacent step to G3 meets an

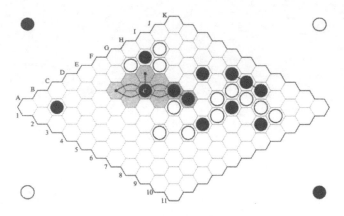

Figure 8.6. Initial steps taken from piece c_3.

adjacent step from chain c_2. These are combined to create another 1-path from chain c_3 to the edge.

These two 1-paths are disjoint and can be consolidated to give a single 0-path from c_3 to the edge. This 0-path is in turn safely connected to chain c_4 by a bridge step, which is safely connected to c_7 by a bridge step, which is safely connected to the bottom right edge by the ladder escape (Figure 8.7).

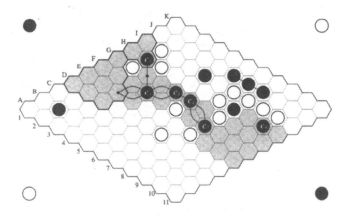

Figure 8.7. The completed spanning path is 0-connected.

Black's best spanning path has been found and is 0-connected. There is no need to search further for additional spanning paths, and the algorithm terminates at this point. The best spanning path was detected very quickly, after only one step from the initial groups.

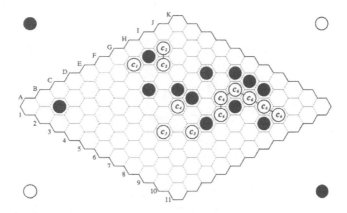

Figure 8.8. White's chains.

8.2.2 Path Evaluation for White

We now turn to the evaluation of White's position on the same board. Given that Black's spanning path is 0-connected we know that White's best spanning path can at best be 2-connected (see Section 4.3.5 and Appendix D.4 for an explanation of this).

White's chains are shown in Figure 8.8. Links between adjacent pieces within groups are shown. These chains and the two White edges make up the initial collection of singleton groups.

Edge templates that safely connect chains c_3 to the bottom left edge and c_6 to the top right edge are shown in Figure 8.9. These are templates IVb and IIIb. Although several edge template steps that terminate on empty points exist, none are relevant to the best spanning path and are not shown here.

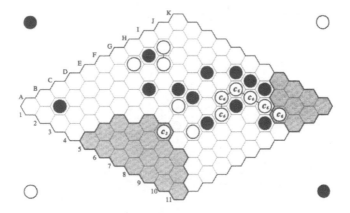

Figure 8.9. Edge templates to safely connected chains c_3 and c_6.

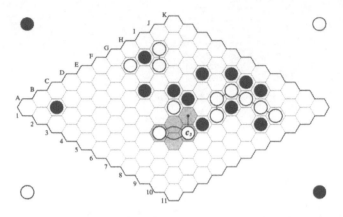

Figure 8.10. Initial steps taken from piece c_5.

Some of the initial steps taken from the singleton groups formed by piece c_5 are shown in Figure 8.10. These include a bridge step to chain c_3 and an adjacent step to F1 with pivot point shown.

A single step was not sufficient to determine White's minimum spanning path. Figure 8.11 shows the step taken from chain $c5$ to F7 extended to point G7 to give a 2-step. This 2-step has two pivot points: an internal one from the previous step, and its terminal point.

This 2-step combines with an adjacent 1-step from c_6 as shown in Figure 8.12 to give a 2-path between c_5 and c_6. This allows the completion of a path between the two edges by path extension, giving a 2-connected spanning path.

This is bad news for White. Can they do any better?

On the same iteration of the algorithm, the alternative path shown in Figure 8.13 is found. It is also only 2-connected, but it is disjoint with the path found previously. Let's call these two paths $best_1$ and $best_2$.

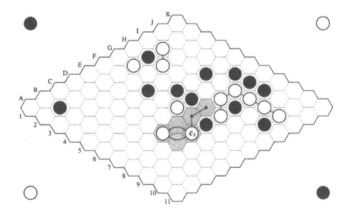

Figure 8.11. The step from c_5 to point F7 is extended to point G7.

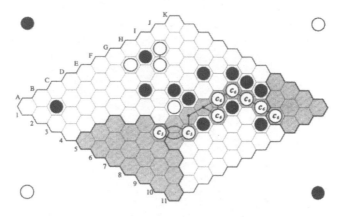

Figure 8.12. White's best spanning path is 2-connected.

It would be convenient for White if $best_1$ and $best_2$ could be consolidated to a single 1-path to improve their connection. After all, these paths are disjoint and have as terminals the same groups:

- $<(e_0, c_3, c_5)\{A5\text{-}11, B5\text{-}10, C5\text{-}8, D6, D8, E7\}>$
- $<(c_6, e_1)\{I10, I11, J9, J10, J11, K8, K9, K10, K11\}>$

where e_0 and e_1 are White's bottom left and top right edges.

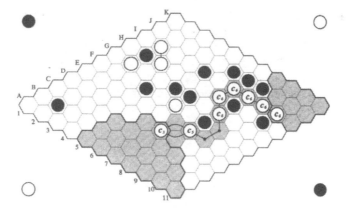

Figure 8.13. An alternative 2-connected spanning path.

However, recall from Section 4.3.4 that only 1-paths may be consolidated. Since the player only has one move in which to answer an intrusion before the opponent is able to again intrude on their next turn, any combination involving 2-paths can not be safely defended (Figure 8.14)).

White has two best spanning paths, each of which is 2-connected.

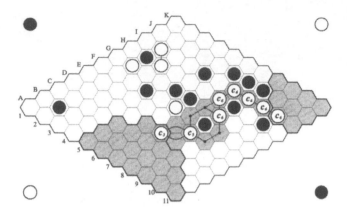

Figure 8.14. This pair of 2-connected paths can not be consolidated

8.2.3 Combined Path Evaluation

It's obvious from the outcome of this pair of board evaluations that Black has won the game, regardless of whose move it is next turn.

The weakest links within Black's spanning path are:

- *the 1-paths consolidated between the top left edge and chain c_ℓ and*

- *the ladder escape from chain c_r.*

Playing at pivot points within these weaker areas will give Black the quickest victory. The 2-connected links within these paths represent the weakest links in White's best connection across the board, which White must attend to as soon as possible.

The regions of overlap between the two players' best spanning paths (as shown in Figure 8.15) indicate key points on the board. In the absence of any better move, it is usually a good idea to play within these regions. However, it's too late for White in this case.

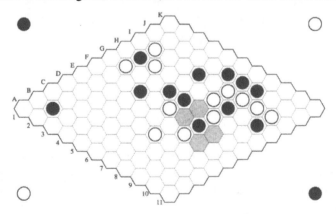

Figure 8.15. Region of overlap between Black's and White's best spanning paths.

A useful estimate of the importance of each point within the best spanning paths can be given by determining the percentage of spanning paths that pass through each point and the degree of their connection. For instance a point with 90% passage would indicate that almost all of a player's best spanning paths channel through it, making it an interesting target.

Given two points of similar value, the point that belongs to the weakest link is the most vulnerable. Hence a 75% point belonging to a 1-step would be more vulnerable than a 75% point belonging to a 0-connected bridge step.

If the most vulnerable points are of equal value and of similar connectivity, pivot points are the most important.

8.3 The Need for Groups

The number of groups generated in a board evaluation will always be at least the number of chains on the board plus the number of player's edges (two). It will generally be much greater than that number, especially for more complex analyses that generate large numbers of paths, as all possible group combinations must be examined until a spanning path is found. The algorithm would therefore be more efficient if path generation depended only on steps taken from individual chains rather than groups. Why are groups necessary in the evaluation process?

Groups provide the means for more complex and accurate development of step combinations than is provided by steps from chains alone. Consider White's position on the board shown in Figure 8.16.

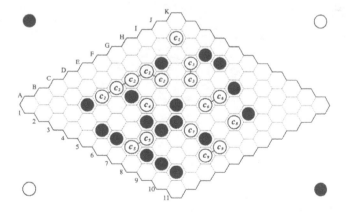

Figure 8.16. White chains with adjacent links shown.

Figure 8.17 shows a minimal spanning path for White based on chains, steps, and paths only. This path is 2-connected, its weak link being the two pivot points between chains c_4 and c_5. There are several other 2-connected spanning paths based on chains, steps, and paths only.

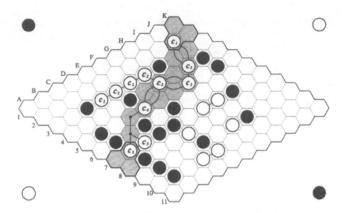

Figure 8.17. One of several 2-connected spanning paths based on chains, steps and paths only.

White's best spanning path can be improved dramatically if steps are taken from groups rather than individual chains.

Figure 8.18 shows the safe group $<(c_2, c_3, c_4, c_6, c_7, c_8, c_9)$ {F4, F5, F7, F8, F9, F10, G5, G6, G8, H4, H6, H8}>. Bridge links connecting the chains within the group are shown.

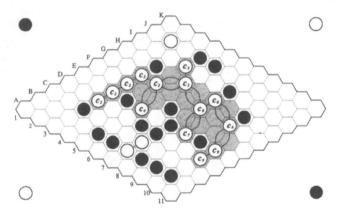

Figure 8.18. Safe group $<(c_2, c_3, c_4, c_6, c_7, c_8, c_9)$ {F4, F5, F7, F8, F9, F10, G5, G6, G8, H4, H6, H8}>.

Steps are now taken from the group as a whole. Figure 8.19 shows the result after the initial collective step is taken from four chains within this group. These are all 1-steps. Only steps eventually contributing to the best spanning path are shown for clarity.

Figure 8.20 shows the result after a second step is taken from the group. These steps were originally 2-steps but were consolidated to 1-steps as each pair was found to share a terminal point and not overlap.

Thus steps from the four chains converge to a pair of 1-steps.

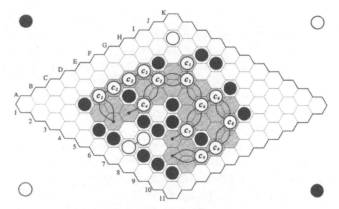

Figure 8.19. First step taken from the group gives four 1-steps.

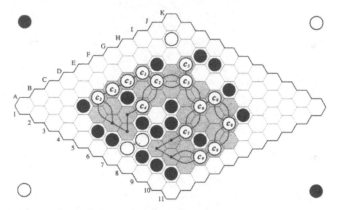

Figure 8.20. Second step taken from the group converges to two 1-steps.

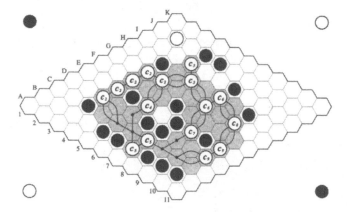

Figure 8.21. Third step taken from the group converges to a 0-path between c_5 and the group.

Figure 8.21 shows the result after another iteration. The existing 1-steps are extended to give a pair of 2-steps, which converge on chain c_5 without interfering with each other, and hence can be consolidated to a single 1-step. As this 1-step terminates on a group $<c_5>$ it forms a 0-path, which is equivalent to the new group $<(c_2, c_3, c_4, c_5, c_6, c_7, c_8, c_9)$ {C4,C5,C6,C8,C9,C10,D4,D5,D8,D9,F4, F5, F7, F8, F9, F10, G5, G6, G8, H4, H6, H8}> shown in Figure 8.22. Pivot points are underlined.

This group combines with template II to each of White's edges to give the 0-connected spanning path shown in Figure 8.22.

This example shows how groups allow a more powerful and accurate measure of connectivity across the board than individual chains can provide. The concept of grouping essentially works to collapse collections of safely connected chains into single units that can be used as a common base for step expansion.

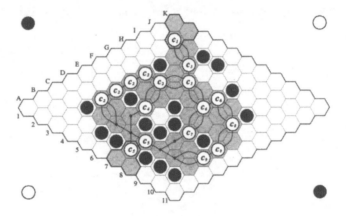

Figure 8.22. The correct best spanning path is 0-connetced.

8.4 Optimizations

A number of optimizations can be applied to implementations of the algorithm to improve its performance. Some of these are discussed below.

Superset steps and paths (steps and paths that are identical to existing steps or paths except that they contain more empty points) should be deleted before being added to the step and path collections. This can be further optimized by not creating superset steps or paths in the first place. This deletion of supersets is valid as their corresponding subsets represent an identical connection, only better, since they:

- *contain fewer empty points in which the opponent can intrude,*

- *are easier to place on a crowded board, and*

- *have fewer possible component parts, which improves complexity.*

As a general rule of thumb, if two groups share an n-path between then, then $n+2$ paths between these groups will very rarely be relevant to the path generation process and can be removed with relative safety. This assumption has proven sound for a variety of test situations but should be applied cautiously as it is not proven to be true. A less sound assumption, but one that improves performance dramatically, is that $n+1$ paths also be removed. This assumption should be applied only if a gross estimate rather than an exact measure of connectivity is required.

If the player's best spanning path is 0-connected the game is won, and the algorithm can terminate on the first such path found. After establishing a winning connection, the computer Hex players may prolong the game unduly by playing at random empty points within the connection. Preferring to play at pivot points within the winning connection will reduce this problem.

Predefined edge templates are used to kickstart path generation from the edges by immediately identifying connected chains. This saves considerable computation over developing the same formations from a series of steps and paths.

A computer Hex implementation may benefit from precomputing known patterns such as the initial set of steps that may be taken from any given singleton group (chain) and storing this information in a table for quick lookup. The maximum chain size to be stored must be selected carefully, however, due to the large number of possible chain configurations (see Appendix E, Polyhexes). The following information would need to be stored for each step:

- *the step's order of connectivity,*

- *its terminal point,*

- *its empty point set, and*

- *pivot points within the empty point set.*

8.5 Features of the Algorithm

The algorithm is excellent at giving a measure of connectivity across the board and detecting when a game is won/lost, but does not in itself indicate exactly where the player should move in order to achieve/avoid this result. However, much useful information is captured by the evaluation:

- *It identifies moves of limited interest (moves outside the connectivity regions of both players' best spanning paths and do not provide ladder escapes) that do not warrant closer investigation.*

- *Within the moves belonging to at least one player's spanning path, it identifies moves of particular interest.*

- *Interesting moves can be approximately ranked in order of priority.*

•   *The iterative step-based approach inherently contains connectivity information that would otherwise only be determined by playing through several lines of play. Thus complete areas of the game tree may be collapsed into a single evaluation.*

•   *Forcing moves can be recognized as opponent's strong links through which a majority of paths flow. If the opponent intrudes in such a link then the player is forced to complete the link if they wish to maintain that set of spanning paths.*

Due to the nature of Hex, experienced players will generally allow each other very limited connection opportunities. This results in a very fast board evaluation with few combinations to be considered. Ironically, poorly played board positions generally require considerably more analysis than expertly played board positions due to the large number of potential but ill-defined connections.

Some current Hex playing programs use adjacent and bridge steps only when measuring connectivity. Without step and path combination or the use of groups, these measurements do not describe the true connectivity across the board.

Summary

Each player's best spanning path across the board can be determined by treating connected chains of their pieces as groups from which steps may be taken iteratively, forming new paths and groups from which further steps may be taken. This process is repeated until the best spanning path across the board is found.

The board evaluation algorithm gives a measure of each player's connectivity and determines whether a game has been won. It also provides key information about the structure of the players' positions, and key points within them that are candidates for good moves.

9

Opening Play

The first few moves of any game establish the connective framework upon which the rest of the game is based. Any mistake at this stage can be catastrophic. Players must strive to make moves that not only threaten to directly form strong connections, but also provide potential for improving their position later in the game. Usually both players will try to create a loosely defined connection between their edges that solidifies as the game progresses.

This chapter examines common opening plays and suggests ways to establish a good foundation upon which to build a solid game. It does not occur towards the start of the book as one might expect, as some understanding of more advanced topics such as ladder handling and path analysis is required to appreciate the full subtlety of opening play.

9.1 Opening and Swapping

The first player has a huge (winning) advantage if allowed to open at a position of their choice. It's been proven that the first player has a winning line if they do not make any mistakes (see Appendix D.2), although that line has not yet been defined.

The following sections examine the best opening move, ways to defend against it, and how to achieve a more balanced game using the swap option.

9.1.1 First Move

Although the precise winning strategy for the opening player has not yet been defined, it's safe to say that the central hexagon F6 is the strongest opening on an 11x11 board. This almost guarantees a win for an experienced player.

The strengths of F6 have been discussed in previous sections. It reduces the largest gap in the opening player's best spanning by half, and is only one step away from connecting to each of the player's edges.

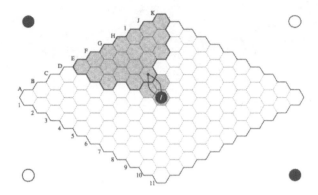

Figure 9.1. F6 is the strongest opening move.

Figure 9.1 shows the single bridge step leading to edge template IVa, giving Black a 1-path to the top left edge. A similar connection on the other side of the board forms a 1-path to the lower right edge.

It is assumed in this chapter that Black has first move for the sake of consistency. This not necessarily the case as there is no universally accepted starting color. The player to move first may be chosen by the toss of a coin, by mutual agreement, or according to the outcome of previous games. It is common for opponents to engage in *dual challenges* (two games of Hex started in parallel, with each player making the first move in one game) to average out the first move advantage.

9.1.2 Defense Against First Move at F6

Given that the opening player Black will take F6 if given the choice, what is White's best defense? White must defend on one side or the other; let's assume that White's intention is to block Black's opening move *1* F6 from the top left edge.

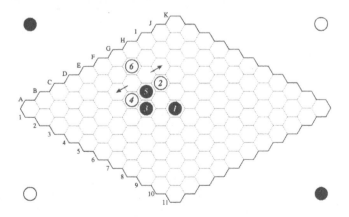

Figure 9.2. The pivot point is not a good defense against F6.

A good way to evaluate a move's worth is to mentally play it through and consider the most damaging sequence of play that may result. White must play inside the shaded region of Figure 9.1 to intrude into Black's best connection. The most promising moves within this region shall be examined in turn.

The obvious place to start is the major pivot in Black's connection at G4. A likely sequence of play is shown in Figure 9.2. If Black responds with bridge *3* E5, then White is wise to stop this advance with move *4* E4. The best that White can hope to achieve is to limit Black to a 3-row ladder as indicated. This is not a particularly good outcome for White.

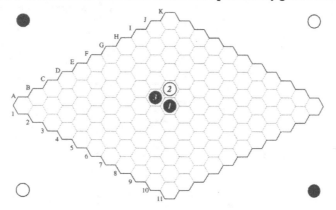

Figure 9.3.   Dreadful defense by White.

Now consider the two vulnerable points in Black's bridge step from F6, namely F5 and G5. Playing *2* G5 is a dreadful move for White; it's far too close to F6 and does not impede White's passage to the edge, as shown in Figure 9.3. Black's move *3* F5 in the bridge's dual point only strengthens their connection. White is cut off from the left side of the board and is in a disastrous position already.

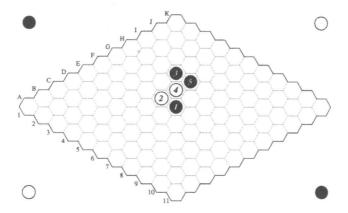

Figure 9.4.   The other vulnerable bridge point F5 is not much better.

Playing in the bridge's other vulnerable point F5 is not much better, as shown in Figure 9.4. Again White's move *2* is far too close to F6. Black's reply *3* H4 more or less forces White's *4* G5 to keep F6 separated, but with *5* H5 Black has cut White off from the right hand side. This is again a bad result for White.

White's defensive move *2* F4 shown in Figure 9.5 is further removed from F6 but is still a poor defense. It's not in Black's direct line of connection, and they are able to make a serious threat with *3* G4. White is forced to play *4* and concede a 2-row ladder, not a good result.

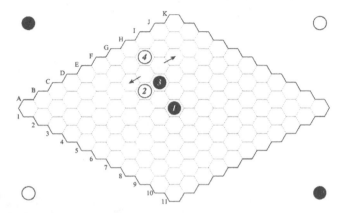

Figure 9.5. Move 2 at F4 allows Black to make a strong threat.

What if White tries to defend more towards the edge? Figure 9.6 shows the White's unfortunate result after playing *2* G2. This point falls within the *don't care* region of template IVc and Black is able to re-establish their connection with *3* F4. They are now in a stronger position. Other defensive moves *2* by White along row 2 are equally bad.

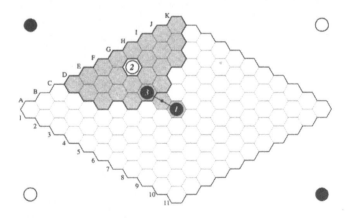

Figure 9.6. G2 falls within edge template IVc.

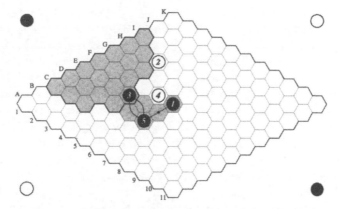

Figure 9.7. A better attempt but still to Black's advantage.

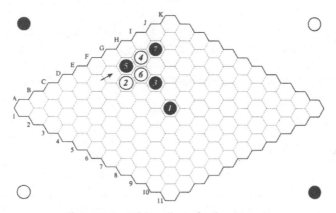

Figure 9.8. White allows forcing move 5.

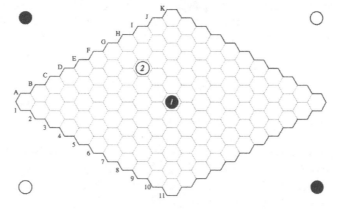

Figure 9.9. Best reply to an opening at F6.

The defense *2* H3 shown in Figure 8.7 is slightly better. This play is midway between the opening piece and the edge, and lies in the classic defense position in relation to F6 (see Sections 3.1.3 and 6.7). However Black can sidestep the defense with *3* E4 and *5* D6 to push a strong connection down the left side.

Figure 9.8 shows a likely sequence of play if White plays *2* F3. Black achieves a threatening direct connection to the edge through pieces *3* and *7*, and along the way is able to play the potentially dangerous forcing move *5* G2. This piece gains territory for Black and provides a potential escape for ladders progressing along row 2 as shown.

White's best defensive move is G3 as shown in Figure 9.9. This move is in the classic defensive position relative to F6 just as move H3 was, but is one step closer to the long diagonal axis of the board. This is to White's advantage as this move reduces the largest gap in their own best spanning path, and successfully impedes Black's *1* F6. The proximity of White's top right edge makes it difficult for Black to force a connection through the gap to the right of piece *2*.

9.1.3 The Swap Option

Although G3 is a reasonable defensive move, Black is still in a superior position and should win the game from this point with good play. F6 is an overwhelmingly strong opening move. Recall from Section 1.2 that an additional rule can be allowed to reduce the first player's advantage. This is the *swap option*, and is described as follows:

•   *The player to move second has the choice of swapping colors, effectively stealing the first player's move.*

The swap option is a way of ensuring that the first player does not make too strong an opening move. It's called the Pie Rule by Wayne Schmittberger in relation the game of Star for obvious reasons: "I cut the pie, you choose the piece you want" [1983].

A number of alternative methods may be used to reduce the first player advantage. For instance, the first player may not be allowed to open on the strong central line or within a certain distance of it, or may be constrained to making the first move within a relatively harmless area of the board. Other strategies include letting the opening player set up an initial position of three or four pieces and letting the opponent decide which side to take. Often two games are played in tandem, one player starting each, to balance out the first player advantage. It is reduced somewhat on larger and even-sided boards, but is still a winning advantage.

The swap option is arguably the fairest and most elegant solution, and makes the first player choose their opening move very carefully. It's still advantageous to keep the first move, so the opening player should play as strong an opening as possible without encouraging the opponent to swap. There is the danger of playing too weak an opening, to which the opponent responds with F6 and takes control of the game.

In many ways the swap option does for Hex what the doubling cube does for Backgammon: it alleviates a fundamental flaw in the game and adds another dimension of strategy [Jacoby and Crawford 1970]. It's recommended that the swap option *always* be allowed between players of reasonably equal skill.

Strategies for handling the swap will now be discussed.

9.1.3.1 When to Swap

If the opening move is F6, the second player should obviously swap. Beyond this simple
rule players must develop their own preferences for determining whether to swap or not.

Figure 9.10 shows estimated swap values for opening moves on the 11x11 board that
experience has proven to be sound. White hexagons indicate opening moves that should
be swapped without hesitation. These moves are too good to ignore and are increasingly
strong towards the central point F6.

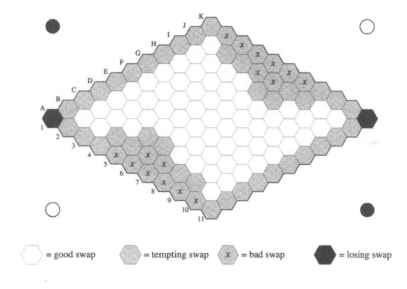

Figure 9.10. Opening moves ranked by swap value.

"Tempting" borderline opening moves (shaded) are not particularly strong moves in
themselves, but are probably worth swapping given that the player will have the added
benefit of first move initiative. Several provide ladder escapes that may prove useful later in
the game, and it's generally to the player's advantage to have an extra piece on the board.

Stronger opponents will use these opening moves to good advantage if they are not
swapped. However against weaker opponents, who are unable to take full advantage of
these borderline openings, the player may be better advised not to swap but to go on the
offensive with a move at or near F6.

Shaded hexagons marked *x* represent substantially weaker borderline openings that
are generally unwise to swap.

The two acute corner hexagons A1 and K11 are losing openings, as shown in Figure
9.11, and should never be swapped. If Black opens in the acute corner with *1* A1 then
White's reply *2* A2 essentially removes this piece from the game. The set of empty points
touching *1* is reduced to the single point *t* and if Black's piece *1* is to play any further part

in the game it would have to connect through t. However such a connection is superfluous as any move at t connects to their edge anyway. White can pretend that their reply *2* A2 is effectively the first move of the game and wrest the first move advantage from Black. White has gained two points of territory t and u in this exchange, whereas Black has gained nothing whatsoever.

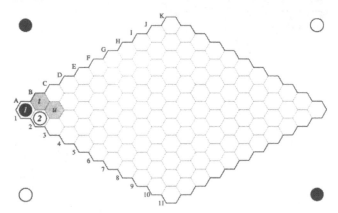

Figure 9.11. A1 is a losing opening.

This exact situation is used by Beck [1969] to prove that the second player has a winning line in Beck's Hex. This is a version of Hex in which the second player dictates where the first move is made. The second player should win if the opening move is made in an acute corner.

The swap values shown in Figure 9.10 are based on experience and have generally proved to be sound, but may not be appropriate for particular playing styles. For example, some players may prefer to establish a strong toehold in the center of the board immediately, and will forego the swap more readily to achieve this. It's up to the individual player to determine the swapping criteria with which they are comfortable.

9.1.3.2 Where to Open Under Threat of Swap

Given that the opponent is likely to swap any good move, where should the first player open? Let's assume that they wish to make the strongest possible move that does not result in a swap.

Moves in the acute corners A1 and K11 can be ruled out immediately. Those moves deemed bad swaps (marked x in Figure 9.10) should also be discarded as it is likely that the player would be left with any of these weak moves.

Figure 9.12 indicates some deceptively strong opening moves (the following arguments also apply to their 180 degree rotated equivalents). *1a* and *1b* each straddle the home areas of both players and are connected to their nearest edges by template II. They also provide ladder escapes for 2-row and 3-row ladders along these edges, and could possibly block the opponent from playing their own ladder escapes.

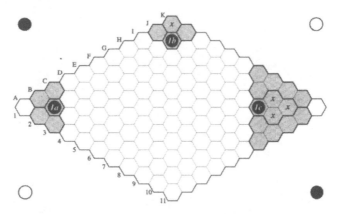

Figure 9.12. Strong opening moves under threat of swap.

The opening *1c* is stronger still. It straddles the home areas of both players, is connected to the nearest two edges by templates IIIa and IIIb, provides ladder escapes for 2-row, 3-row, and 4-row ladders, and could possibly block the opponent from playing their own ladder escapes. It gives the player an immediate connection to the third row from the edge, a strong advantage indeed.

These moves are excellent openings against an intermediate player who may not realize their potential and decline the swap. However, they are far too strong to play against an experienced opponent, who will most likely swap.

Against a stronger opponent it is usually wise to choose opening moves from the "tempting" category. A subset of these moves is shown in Figure 9.13.

These are relatively safe openings, and no real harm will be done if the opponent chooses not to swap. A few moves belonging to this set, marked *o*, stand out as optimal openings. These moves form connections to their nearest edges, provide ladder escapes

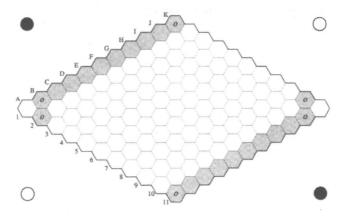

Figure 9.13. Safe opening moves under threat of swap (optimal moves are marked *o*).

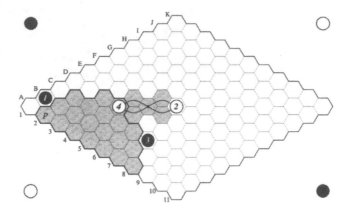

Figure 9.14. A2 is a stronger safe opening.

for the player, and impede opponent's ladders. Among these safe openings A2 and K10 are
the best choices as they provide both 2-ladder and 3-ladder escapes.

The opening sequence shown in Figure 9.14 further illustrates the usefulness of opening A2. Had Black played opening move *1* B1 instead, White's *4* D4 would be safely
connected to the edge by template IVb and put them in a strong position. By opening one
point away at *p*, Black could have prevented this template forming.

A good rule of thumb when choosing an opening move under threat of swap is *the
stronger the opponent, the safer the opening move should be*.

9.2   Common Opening Strategies

The following sections discuss some of the more common opening strategies. These discussions are based upon the assumptions that:

- *Black plays first, and makes a safe but strong opening,*

- *the swap option is allowed,*

- *White takes the central hexagon F6 or the nearest suitable point.*

A typical opening situation is illustrated in Figure 9.15. For the sake of argument let's
assume that these are the first three moves made in each example. Most games begin with
a similar sequence, then diverge dramatically. This configuration allows the basic principles to be demonstrated and provides a common ground for comparison between common opening situations.

Move *1* A2 is a good choice for first move as it is something of a saddle point: strong
enough to keep as it provides an escape piece for 2-row and 3-row ladders, but not strong
enough to cause an automatic swap. An experienced player would probably swap it, however.

Note that if Black plays first then piece *1* will remain Black regardless of whether White
swaps or not; the order of play is swapped, not the piece or the direction of play.

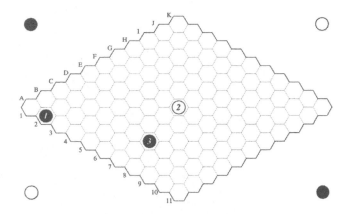

Figure 9.15. A common opening formation: open, swap, F6, and block.

White would most likely take this opportunity to secure the central hexagon with *2* F6. Black can impede their connection to the bottom left edge with the classic blocking move *3* C7. C7 is preferable to the corresponding move I5 on the opposite side of the board for a number of reasons. It keeps Black's pieces within some reasonable distance from each other and potentially able to threaten a connection in a few moves' time. A move on the opposite side of the board would split Black's defense down the middle. Pieces working together are usually stronger than pieces working in isolation.

In general, it is better to *play defensive move 3 on the same side of opening move 1*, if it occurs near an acute corner. However, if the opening move is made near an obtuse corner, it's usually better to play the blocking move *3* on the *opposite* side to avoid overcrowding in that region.

Where should White play next? The following sections examine some promising options based on this common starting sequence and White's choice of move *4*.

9.2.1 Spread From Center

Move *4* D4 shown in Figure 9.16 is a good attacking play by White, as it:

- *is connected to piece 1 F6 by a 1-step,*

- *is connected to the edge by several 1-steps, and*

- *impedes the connection of Black's piece 3 to the top left edge.*

This move strengthens the weakest link in White's best spanning path, and spreads their connection evenly from their strong foothold at F6. Piece *4* is *almost* connected to the bottom left edge by edge template IVb but is blocked by piece *1*, another reason that A2 is an excellent choice for opening move. Nevertheless, piece *4* has many possible paths to the edge and will be difficult to block.

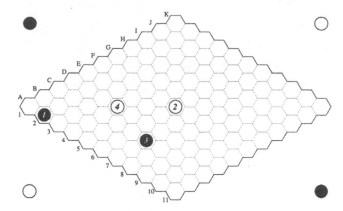

Figure 9.16.   Move **4** D4 is a strong development by White that spreads their connection from the center.

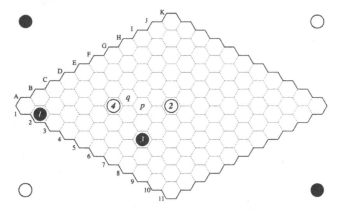

Figure 9.17.   Points *p* and *q* are Black's best defense in this situation.

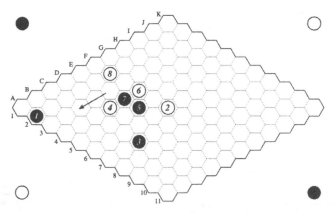

Figure 9.18.   A typical development from point p shows the usefulness of **1** A2.

Playing move *5* at either points *p* or *q*, as shown in Figure 9.17, are both good defensive choices for Black. Move *p* is slightly stronger and is generally used more often, but there is little to choose between the two.

Figure 9.18 shows one possible extension of play from point *p*. White's attempt to stop Black's progress with *6* F4 yields a 3-row ladder for Black, which escapes at piece *1*. Other variations are possible but too numerous to list here, and belong more to the early/middle stages of the game.

Now consider the case if White had instead played move *4* on the opposite side of the board, for instance *4* H5 as shown in Figure 9.19. In many ways this is a good move for White as it safely connects their piece *2* to the top left edge via a bridge and template IVb. However, this is generally a bad idea as Black now has the opportunity of playing at points *r* or *s* to block White and set up a potentially threatening connection.

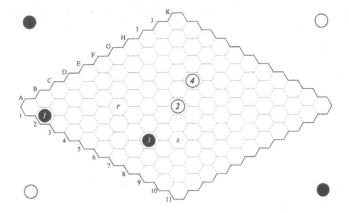

Figure 9.19. White's move *4* on the opposite side is weaker.

Figure 9.20 shows another variation on White's move *4*. *4* E4 is only one point removed from *4* D4 but is arguably weaker. David Boll [1994] recommends the use of *4* E4 as it both forms a loose connection with White piece *2* and is in the classic defensive position against Black piece *3*. However, this does not take into account the presence of opening piece *1* which changes the situation somewhat. E4 is indeed more solidly connected to central piece *2* but it is also one column further removed from the bottom left edge and less threatening as a ladder or connection.

9.2.2 Push To Edge

Another common starting sequence is shown in Figure 9.21. Here White pushes directly towards the edge that Black is defending with a bridge move.

If Black makes the obvious blocking move *5* C8, then White can follow up with the attacking *6* D5, as shown in Figure 9.22. Piece *6* is now almost guaranteed a connection to the edge.

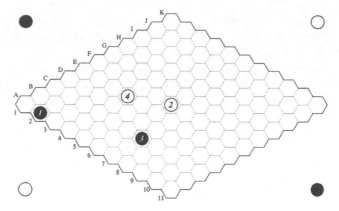

Figure 9.20. Variation *4* E4 is weakened by the presence of opening move *1* A2.

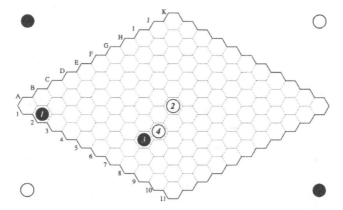

Figure 9.21. White pushes directly towards the edge with a bridge move.

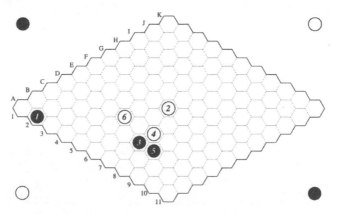

Figure 9.22. White sets up a tangible threat with move *6*.

The best that Black can do here is concede a 2-row ladder along column B, giving White the opportunity to play escape piece *n* at some time in the future, as shown in Figure 9.23. If Black chooses to block this ladder then White is free to push across the board from *n*. This is a dangerous attack as it's also very close to connecting around the corner with pieces *2* and *4*.

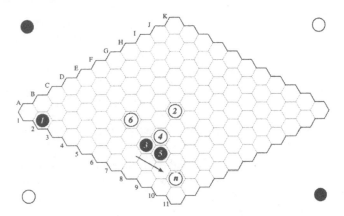

Figure 9.23. Danger of a ladder escape via a future piece *n*.

Instead of *5* C8, a less obvious but better defensive move by Black might have been *5* B8 as illustrated in Figure 9.24. This stops the bridge move *4* from directly connecting to the edge next turn, and also blocks any potential ladders along column B.

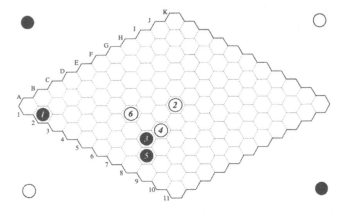

Figure 9.24. A better defensive move *5* by Black cuts of the potential ladder escape.

A path analysis of this improved sequence is shown in Figure 9.25. It can be seen that pieces *2* and *4* are connected by two 1-paths, but that these paths are not disjoint. They overlap at the vulnerable hexagon E6. This looks like a good place for Black to play their next move.

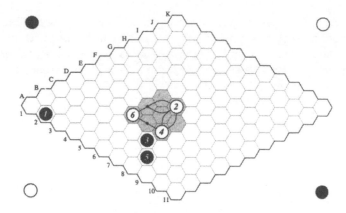

Figure 9.25. Path analysis between pieces *2* and *4* reveals a vulnerable point at E6.

Move *7* E6 (Figure 9.26) has the effect of splitting White's attack, and they must now choose which line to pursue. Two points *s* and *t* are the most likely moves on each line. Again, there are too many possible extensions of this play to consider in further detail; suffice to say that neither *s* nor *t* is the conclusively better move.

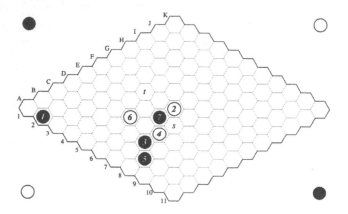

Figure 9.26. White's best reply is at point *s* or *t*.

9.2.3 Push From Edge

In contrast to the previous example, another common opening strategy is to play an edge template move that pushes *from* the edge towards the central piece *2* F6. Two points that achieve this *p* and *q* are shown in illustrated in Figure 9.27.

Playing at point *p* with move *4* C8 as shown in Figure 9.28 is recommended by David Boll [1994]; however, this threat can be weakened with the template intrusion *5* B8 which also foils it as a potential ladder escape.

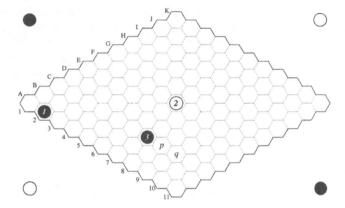

Figure 9.27. Points *p* and *q* push from the edge towards F6.

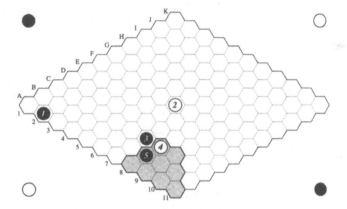

Figure 9.28. Intrusion *5* limits the usefulness of move *4* at *p*.

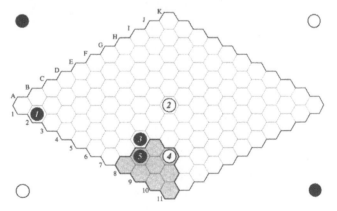

Figure 9.29. The same intrusion *5* also limits the usefulness of move *4* at *q*.

If White instead plays at point *q* with move *4* C9, Black is able to play the same defensive move *5* B8 with the same result, as shown in Figure 9.29. Pushing in from points *p* or *q* is not that strong an opening play.

White's other option is to push towards the center on the other side of piece *3*, perhaps with move *4* C5 as illustrated in Figure 9.30. This does not work very well either, as piece *4* is well within Black's home area and Black is able to cut it off with *5* D5.

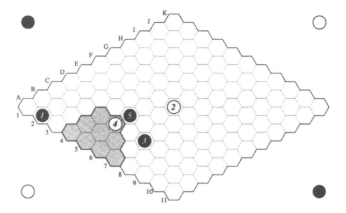

Figure 9.30. Pushing from the edge does not work very well on the other side of piece *3* either.

Move *6* C8 may look promising for White as *4* provides a ready ladder escape (Figure 9.31), but again Black can foil this play with *7* B8.

9.2.4 Block the Opponent

White may choose to block Black's defense with a move such as *4* D8 shown in Figure 9.32. This move looks strong at first but does not lead to good continuations of play for White.

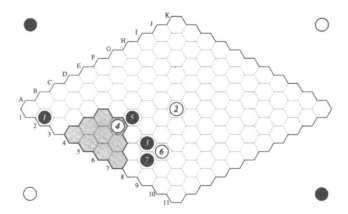

Figure 9.31. White's move *6* looks promising but can be easily foiled.

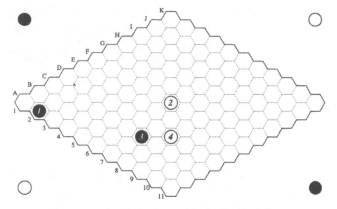

Figure 9.32. White tries to block Black's defense.

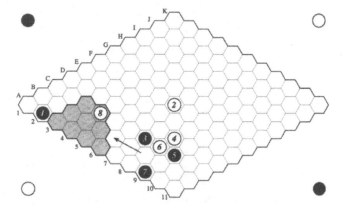

Figure 9.33. Black's reply *5* C9 yields a 3-row ladder for White.

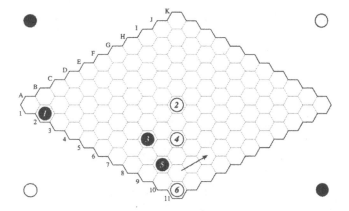

Figure 9.34. Black can force a 2-row ladder if they defend properly.

Black's obvious reply *5* C9 yields a 3-row ladder for White along column C as shown in Figure 9.33. This gives White the opportunity of playing a potential escape piece such as *8* C4 which poses a serious threat, a situation which Black should try to avoid.

Black can solve this problem by blocking White with reply *5* B9 or a similar blocking move (Figure 9.34). Black is able to force a 2-row ladder along row 10 regardless of White's defense; hence, *4* D8 is not a particularly strong move for White.

A better blocking move by White is *4* D5 as shown in Figure 9.35. If Black attempts to stop this connection to the edge directly with move *5* C5 (Figure 9.36), White is able to push around Black's defense to threaten the middle of the board with a move such as *10* D8.

If Black instead tries to separate the two White pieces *2* and *4* with moves such as *5* E6 (Figure 9.37) or *5* E5 (Figure 9.38) then White is again able force the situation to their advantage, as shown.

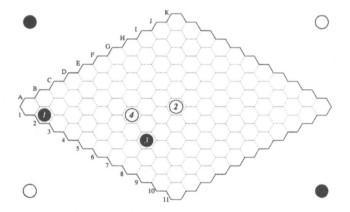

Figure 9.35. Black has difficulty stopping this connection to the edge.

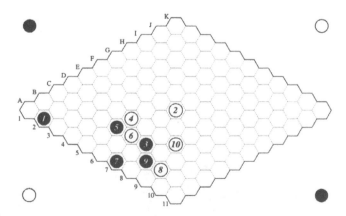

Figure 9.36. A better blocking move 4 by White.

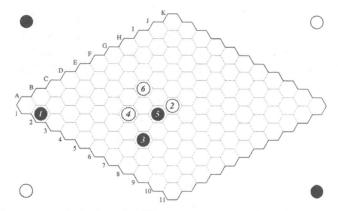

Figure 9.37. Black also has difficulty separating White's pieces 2 and 4.

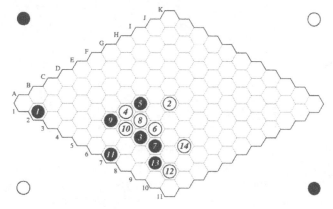

Figure 9.38. White can force a strong position.

9.3   Adapt to the Situation

The above analyses have been based on the A2-F6-C7 starting sequence, which will obviously not always be the case. Some variation in opening play is a good thing as it stops the player from falling into a rut and keeps the opponent guessing. If the player consistently loses against a particular opponent, a change of opening strategy is one of the first remedies to try.

Consider the board situation shown in Figure 9.39. The opening move *1* D4 is far too strong and the opponent has no hesitation in swapping. In fact, it's so strong as to impinge upon the central hexagon and White's reply *2* F6 is not the best. E7 would probably have been a better move, as it:

- *improves the weakest link in White's best connection,*

- *is on the short diagonal, and*

- *impedes piece 1 more than F6 does.*

Black is now able to attack with *3* D7 rather than the classic defense C7 and put themselves in a strong position. White's initial impulse may be to separate pieces *1* and *3* by playing at vulnerable points *p* or *q*, but some analysis reveals both of these moves to be disastrous. White must try to block Black's connection with a move in the vicinity of *r*, and is already on the back foot.

Black has adapted their play to the situation and come out ahead.

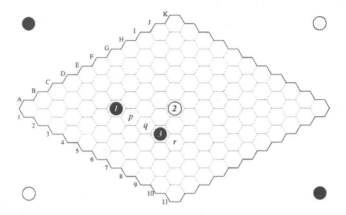

Figure 9.39. Black has adapted to force a strong position.

Of course there is always the chance of adapting badly to a situation, as shown in Figure 9.40. Black plays the weak defensive move *3* I5 on the side of the board opposite their opening piece *1*, to which White replies with *4* H8. Black now breaks off their defense on the right side of the board to play the other classic defensive move *5* C7.

White's reply *6* D4 puts them in a commanding position: they now have a 2-connected spanning path, while Black's best spanning path is 4-connected. White's pieces are well

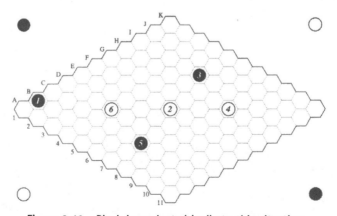

Figure 9.40. Black has adapted badly to this situation.

spread across the board and the weakest links in their spanning path only 1-connected, while Black's defense is split down the middle and in tatters.

Black's errors on this occasion were to:

- *split their defense with move 3 on the opposite side of the board to piece 1, and*

- *ignore the threat posed by piece 4. Recall that it is generally wise to respond to the opponent's last move (Section 3.2.3).*

9.4 Even-Sided Boards

Opening strategy on even-sided boards is substantially different as there are two optimal points, each closer to a different edge. This is shown on the 14x14 board in Figure 9.41. Examples of other even-sided boards from 4x4 to 26x26 are provided in Appendix F.

The proximity of the edges on smaller boards such as 10x10 dramatically changes the game. It's extremely difficult to stop a piece played on one of the two central points from connecting to the nearest edge. This leads to a shorter and less interesting game, so we shall consider only boards 11x11 and larger.

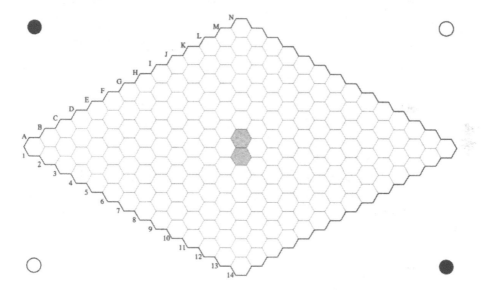

Figure 9.41. The two strongest points on the 14x14 board.

As the board size increases so does distance from the central hexagon(s) to the nearest edge(s), and the less critical the opening play becomes. This added distance gives both players a chance of recovery if they make an error. Play on smaller boards is less forgiving.

Even-sided boards require a modified opening strategy; if the opponent occupies either of the central hexagons on their second move, the player should *defend on the side with the greatest distance to the edge.* This is shown in Figure 9.42. Black exploits the fact that White's *2* G8 is one point closer to their left edge than to their right one by playing move *3* J7, which interferes with the weakest link in White's strongest connection at an early stage of the game. This is more important than playing on the same side as the opening move.

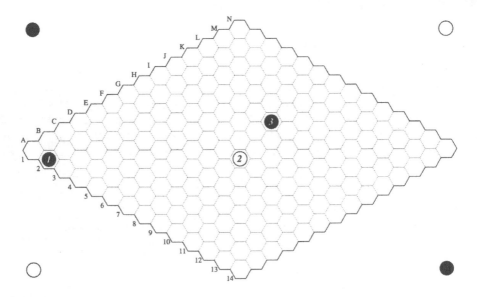

Figure 9.42. Sensible defense *3* by Black.

John Tromp points out the interesting fact that Black wins with third move *3* E6 on the more limited 10x10 board, as illustrated in Figure 9.43. Black has already formed a dominant position against which White has no known defense. The value of the equivalent move *3* on boards 11x11 and larger is questionable but worthy of further investigation.

Summary

The central hexagon F6 is the strongest point on the board and the best opening move. The first player should win if they are allowed to take F6. To counteract this first player advantage, it is recommended that the second always be given the option to swap the first move. This balances out the game and adds a new dimension to Hex strategy.

The player moving first should strive to keep the first move. This means that they must play a move strong enough to give them an advantage, but not strong enough that the opponent will swap it. A move close to one of the opening player's edges satisfies these criteria.

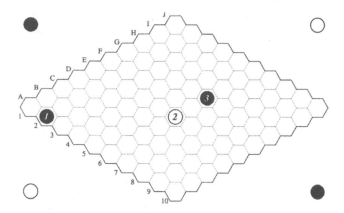

Figure 9.43. Black wins with move *3* E6 on the 10x10 board.

The second player will usually take the central hexagon F6 on their second move. It is wise for the first player to play a classical defensive move on the same side of the board as their opening piece, unless they are playing on an even-sided board.

Openings on even-sided boards require slightly different handling due to the fact that there are two central points, each of which is closer to one edge. The importance of the opening play decreases as board size increases.

10

Strategy III: Advanced

Concepts introduced in previous chapters are drawn together to describe subtle and more involved strategies that help round out a good game. Examples from actual gameplay situations are used to illustrate these advanced points where appropriate.

An understanding of path analysis (Chapters 4 and 8) and ladder handling (Chapter 7) is assumed for many of these points of strategy.

10.1 Multiple Threats Per Move

Just as alternative paths between pieces improve their connectivity, so too do multiple threats increase the usefulness of a move. This is most obvious in forking moves which threaten to complete two or more connections directly, but also applies to less tangible goals that may accumulate to make a move exceptional. A move that combines a number of subtle threats but no single dominant attack is an example of a *zwischenzug* or *intermediate move*.

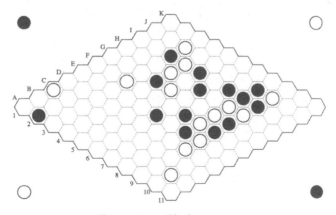

Figure 10.1. Black to move.

This point of strategy is also relevant to the game of Y, where it is described as the concept of Double Trouble [Schensted and Titus 1975].

10.1.1 Combined Threats

The first example demonstrates how less obvious threats may combine to make a good move. Consider the board situation illustrated in Figure 10.1, with Black to play.

Black's position can be better understood following the path analysis shown in Figure 10.2. Piece *17* is safely connected to the lower right edge, even if White blocks it to form a ladder. The ladder can escape along the direction shown using the two adjacent pieces at F9 and G9.

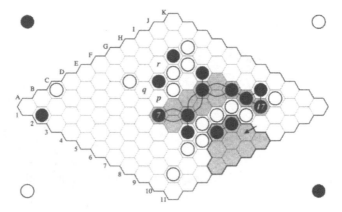

Figure 10.2. Path analysis shows Black to be in a strong position. Moves p, q, and r look most promising.

Piece *17* is in turn safely connected to piece *7* by the path shown. Black's task is now to connect piece *7* to the top left edge. Of all available moves, those at points *p*, *q*, and *r* look the most promising. These will be investigated in turn.

10.1.1.1 Scenario 1: Move p

If Black chooses to play at point *p* with move *29* F5 as shown in Figure 10.3, White's best response is to block the immediate connection with *30* H2. If Black tries to force the connection with *31* H3, then White can complete the block with *32* J1. White is now safely connected to the top right edge, and Black will have to try another tack.

Black's forcing move *33* G2 prepares a potential ladder escape (Figure 10.4). However, White is able to force a ladder escape of their own with moves *36-42* that gives them the win.

10.1.1.2 Scenario 2: Move q

The sequence of play following Black's move *29* F4 at point *q* is shown in Figure 10.5. The first few moves of this sequence are much the same as that for moving at *p*, but in this case the first move is a bridge that allows White to intrude with forcing move *30* E5 that soon impedes Black's progress.

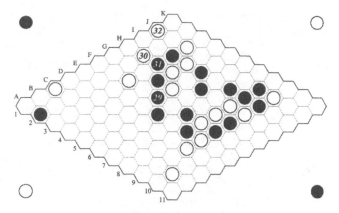

Figure 10.3. Black plays move *29* F5 at point *p*...

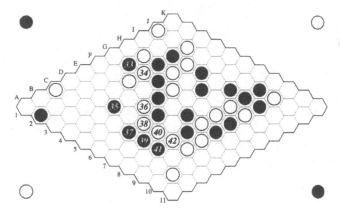

Figure 10.4. ... and loses.

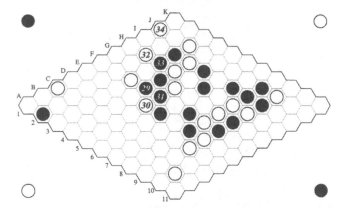

Figure 10.5. Black plays move *29* F4 at point *q*...

The end result is much same as moving at point *p*. White forces a ladder escape with moves *38-44* to win the game.

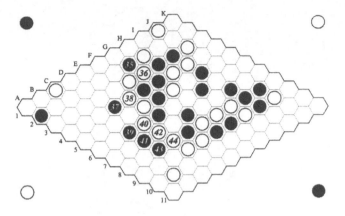

Figure 10.6. ... and loses.

10.1.1.3 Scenario 3: Move r

Now consider the situation if Black plays *29* H3 at the less obvious point *r*, as shown in Figure 10.7. White's *30* F5 is a forced reply. Black can push further with forcing move *31* E5 to which White must reply *32* F4. Move *33* D4 now puts Black in a very commanding position to connect to the top left edge and win the game. White has only two practical defensive moves in this situation: points *s* and *t*.

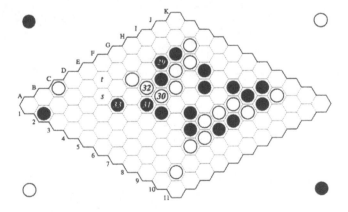

Figure 10.7. Black plays move *29* H3 at point *r* and forces a strong attack.

Notice that piece *29* and its adjacent neighbors provide Black with a ladder escape for a ladder along row 2. If White chooses to follow the defense based on a move at point *s*, then Black can use this to their advantage by forcing a ladder with moves *35-39* as shown in Figure 10.8 to win the game.

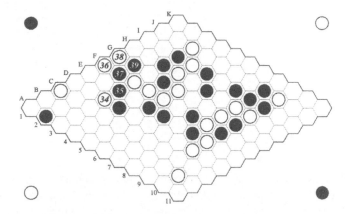

Figure 10.8. Black defeats the line of defense from point *s*.

White's only hope is therefore to play at point *t* with move *34* E2. However, Black can also defeat this defense by playing forcing move *35* C3 followed by *37* B2, which connects with their opening piece at A2.

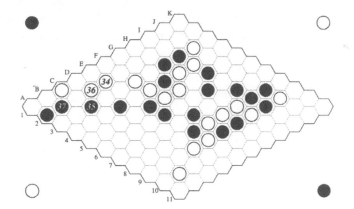

Figure 10.9. Black defeats the other line of defense from point *t*.

Black can force a win with move *29* at point *r*, but will lose if they move at points *p* or *q*. What makes *r* such a strong move? Consider the merits of each move.

Move *29* at point *p*:

- *threatens a direct connection to the edge.*

Move *29* at point *q*:

- *threatens a direct connection to the edge,*

- *gains territory, and*

- *exposes forcing move E5 to White.*

Move *29* at point *r*:

- *threatens a direct connection to the edge,*

- *gains territory,*

- *allows Black forcing move E5 which gains further territory, and*

- *provides a ladder escape.*

From this analysis it can be seen that moving at point *r* achieves a greater number of threats and goals, direct or implied, than the other two choices. The benefit achieved by point *q*'s gained territory is offset by the forcing move that it exposes to White.

As the opponent can only reply once this turn, they must decide which of the multiple threats is most pressing and is to be answered. For this reason, multiple threats within a move will only be effective if they are disjoint and cannot be defeated by a single reply. As a corollary to this point, players should consider the potential paths converging at the point at which they wish to play. The greater the number of *non-overlapping convergent paths*, the greater the likelihood that this point will play an important part in the game.

A count of the threats made by a potential move does not in itself decide its worthiness. However, it does give an indication of the move's potential usefulness, and is a good way to spot killer moves quickly.

Players should *strive to make every move achieve at least two parallel goals or threats, direct or implied.*

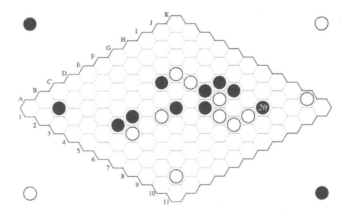

Figure 10.10. White to reply.

10.1.2 Veiled Threats

Given that it's beneficial for a move to contain as many threats as possible, the player can optimize this strategy by choosing moves that hide implied threats behind more obvious ones. There is the chance that the opponent will blindly reply to the greater threat and fail to notice or block the more subtle one, which may return to hurt them later in the game.

A case in point is shown in Figure 10.10, where Black has just played *20* I9. The path analysis shown in Figure 10.11 reveals piece *20* to be safely connected to the lower right edge, and only one step from connecting to the upper right edge. Black's best spanning path contains the glaringly obvious vulnerable point I8.

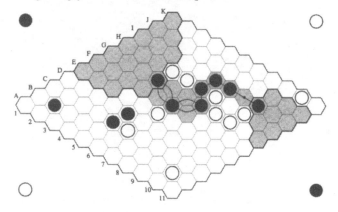

Figure 10.11. Path analysis reveals a particularly vulnerable point.

If White chooses to block the obvious connection with *21* B8, then Black can force a ladder along row 10 with moves *22* E8 and *24* D9, as illustrated in Figure 10.12. Black can now force the win by playing this ladder to its escapes, which was the subtle threat posed

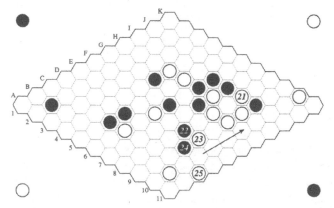

Figure 10.12. White blocks the obvious threat, but falls victim to the more subtle one.

by move *20* I9. It's too late for White to foil this escape as they have been forced to reply to moves *24* onwards.

A much better defensive move *21* by White would have been to play at point *q* as shown in Figure 10.13. This would have interfered with both Black's best spanning path and the potential ladder escape, intruding into a vulnerable point where both threats overlap and defusing the situation.

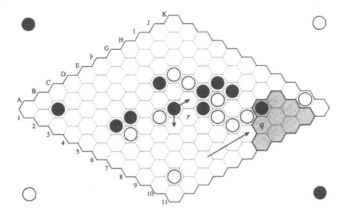

Figure 10.13. Black to move.

Point *r* would also have been a reasonable choice for move *21* by White. This is a more attacking move and forces Black to commit to either of the two lines indicated, weakening the concerted Black attack. Recall from Section 6.7 that it's usually beneficial to reduce the opponent's alternatives.

10.2 Don't Provide Forcing Moves

Forcing moves allow a player to dictate play while maneuvering the opponent into a disadvantageous position, as described in Section 6.3. The best way to avoid being on the receiving end of forcing moves is to *leave as few opportunities for forcing moves available to the opponent as possible*. If it's not practical to avoid leaving a forcing move, then the player should strive to leave the least dangerous forcing move possible.

10.2.1 Concede the Least Dangerous Forcing Move

Consider the board situation illustrated in Figure 10.14, where White has played weakly and is being punished. Their move *6* F10 is especially poor, and neither seriously threatens Black's connection to the edge nor promotes White's connection. Black is now in a position to capitalize on the situation by making a very strong connection with move *7* at either point *p* or *q*. Black should then be able to force a ladder escape to the top left edge via piece *1*, and should not have too many difficulties forcing the connection to the bottom right edge from piece *5*. Which move is best, *p* or *q*?

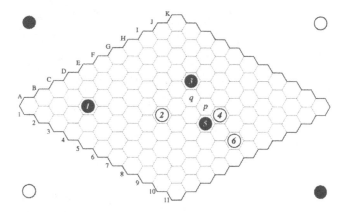

Figure 10.14. Black can make a strong connection by moving at point *p* or *q*. Which is best?

Figure 10.15 illustrates the situation if Black chooses to play move *7* G7 at point *p*, creating a bridge connection between pieces *3* and *7*. White is able to exploit the situation by playing at H6 (shaded), to which Black is forced to reply with G6 to maintain their connection.

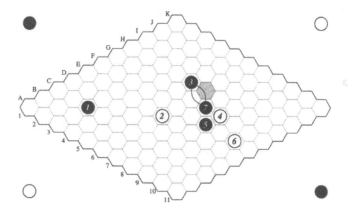

Figure 10.15. Playing *7* G7 at point *p* presents White with a potentially dangerous forcing move.

The situation resulting from this play is shown below in Figure 10.16. This is to White's advantage as they establish a piece *8* on the other side of Black's connection. This piece gains White two points of territory (I5 and I6, shaded), and is exposed to the top right edge and so threatens to provide a ladder escape if ignored. Black has played *9* G6 to reconnect the threatened ladder—a much better move would have been I5, which foils White's potential ladder escape for ladders coming down column I.

Now consider the situation if Black had chosen instead to play *7* G6 at point *q*, as shown in Figure 10.17. In this case, the bridge connection occurs between pieces *5* and *7*.

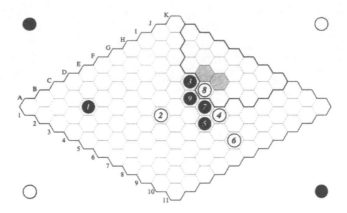

Figure 10.16. White's forcing move **8** H6 looks threatening.

White is not able to exploit this bridge to the same extent; playing in either of the bridge's vulnerable points F7 or G7 does not benefit White in any tangible way.

Although moving at point *q* does technically present just as many forcing moves as the move at point *p*, one for each vulnerable point within the bridge, it is clear that the forcing move presented by the first case is less advantageous for Black. Move *q* is the best choice.

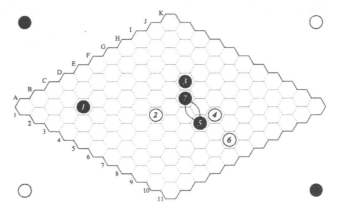

Figure 10.17. Playing **7** G6 at *q* presents White with harmless forcing moves.

A good rule of thumb is to *always assume the worst*. If the player makes a vulnerable move, they must assume that the opponent will see it and exploit it. This assumption forms the basis of most computerized game search algorithms today. It's always a good idea before playing a move to assess the most damaging reply that the opponent can make. This may mean foregoing a clever attacking move in order to play a more solid defensive one, but as pointed out previously there is really little distinction between good attack and good defense in Hex.

10.2.2 Recognize Safe Forcing Moves

Potential forcing moves occur whenever the player makes a non-adjacent template move. In practice this means either:

- *a bridge, or*

- *an edge template.*

Unfortunately, these are also the most useful plays to make in order to develop a connection across the board. Sometimes the player must balance the danger of providing possible forcing moves with the attacking potential of an extended template. As discussed in the last section, the player must learn to recognize when it's safe to concede a potential forcing move.

In general, *a forcing move is benign if it gains no territory for the opponent.*

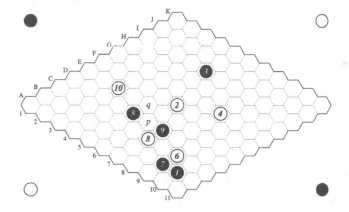

Figure 10.18. Black would like to strengthen their connection by playing move *11* at point *p* or *q*.

Consider the board situation shown in Figure 10.18. Black is developing a reasonable connection across the board and would like to strengthen it by joining pieces *5* and *9* with a move at either point *p* or *q*. Which is the better move?

If Black plays move *11* D6 at point *p* they make a solid connection between pieces *5* and *9* that involves adjacent moves only, and hence leaves no vulnerable points or forcing moves available to White (Figure 10.19).

If Black plays move *11* E5 at point *q* they provide two forcing moves (D6 and E6, shaded in Figure 10.20). However, these forcing moves are benign and gain no territory for White. If White intrudes into the bridge and Black is forced to reply, the situation remains essentially unchanged.

Now consider Black's piece *11*; not only does it connect the pieces *5* and *9*, but it is an attacking play in its own right, threatening to connect to the top left edge to the right of White's blocking piece *10*. The adjacent move *5* threatens to connect to the edge on the left side of *10*. Black is now in a strong position.

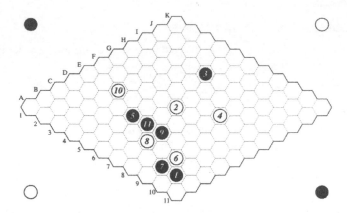

Figure 10.19. Black plays *11* D6 at point *p*, not conceding any forcing moves.

The player should not be afraid to *concede benign forcing moves in exchange for an improved connection.*

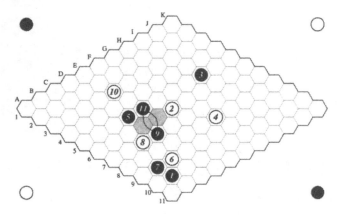

Figure 10.20. Black plays *11* E5 at *q*, conceding two forcing moves but a stronger position.

10.3 Ladder Handling

The mechanics of ladders, ladder escapes, and ladder escape foils were introduced in Chapter 7. This section discusses these techniques in further detail, and demonstrates their application during the game.

10.3.1 Choose the Best Escape

Figure 10.21 shows a board position with two potential ladders indicated for White. Either ladder gives White the win.

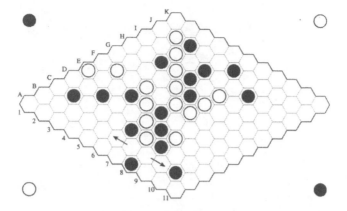

Figure 10.21. Potential ladders for White.

The first winning move is shown in Figure 10.22. White's *32* B9 is a forking ladder escape; it is both a ladder escape template and disjoint forcing move to which Black is forced to reply *33* C9. White can then reply *34* B8 to complete a short ladder and win the game.

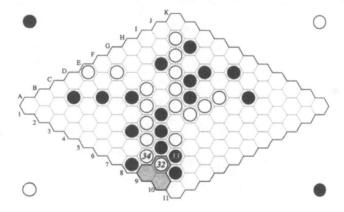

Figure 10.22. The right ladder gives the win...

White can also play *32* C4 to win the game, as shown in Figure 10.23. Black's obvious line of defense is unable to stop the win.

Black may attempt the ladder escape foil following move *32* C4 as shown in Figure 10.24. Defensive move *33* C2 intrudes on the ladder escape's template, and is an adjacent move from a neighboring piece, which was shown in Section 7.5.1 to be a necessary condition for a successful ladder escape foil. Unfortunately, this defensive maneuver occurs on the wrong side of the escape piece and White is still able to make good their escape and win the game. *33* B5 and *33* C5 do not work as ladder escape foils in this case either.

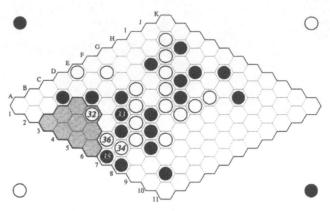

Figure 10.23.   ... as does the left ladder.

The player should always keep an eye out for alternative ladder escape forks and other killer moves, and not just play the first one found. Although the second scenario described above still leads to a win for White, the first scenario is preferable as it leads to a quicker victory with less room for error or unpleasant defensive surprises from the opponent.

The player should *always look for alternatives (even for winning moves) and take the most decisive one available.*

This example is drawn directly from Annotated Sample Game 11.3. Amazingly, Black wins from the position shown in Figure 10.21! White makes a few further inconsequential moves, then resigns without noticing at least two winning plays. The player should *always* be on the lookout for such winning moves, especially before resigning.

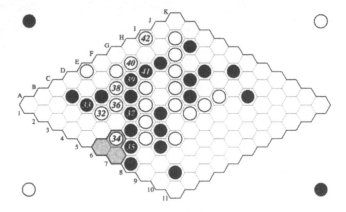

Figure 10.24. A better defense by Black, but White still wins.

10.3.2 Foil Escapes if Possible

Games are usually won and lost on the players' handling of ladders. A good ladder escape is a potential game winner and should not be ignored.

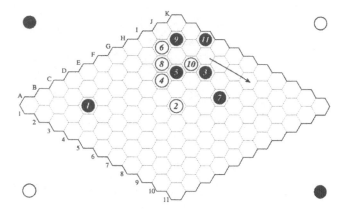

Figure 10.25. A potential ladder.

Figure 10.25 shows a situation where White is threatening to push a ladder down column J. In preparation of the ladder, White attempts to place an escape piece with *12* I6, as shown in Figure 10.26.

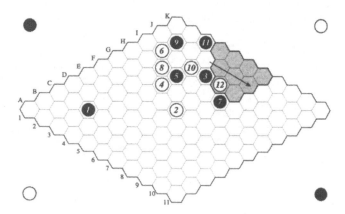

Figure 10.26. An attempted ladder escape...

This is not a very successful attempt at a ladder escape. Black is able to block it with *13* J5, which:

- *intrudes on the ladder escape template,*

- *intrudes on the projected ladder path, and*

- *is adjacent to a nearby defensive piece (3).*

These are the conditions for a successful ladder escape foil, as shown in Figure 10.27.

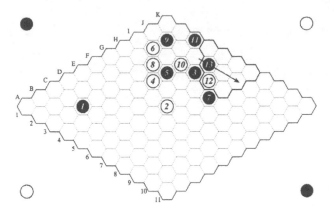

Figure 10.27. ... foiled!

Black might have preferred to delay playing the foiling move *13* J5, secure in the knowledge that they have a blocking move when required and hoping that White overlooks the foil and comes to mistakenly rely on the ladder's success. However, the ladder has already formed and even a delay of a single move would be disastrous for Black. They are obliged to play the blocking move immediately.

10.3.3 Exploit Existing Pieces

Ladder escape templates (Section 7.3) aren't the only way to guarantee a ladder's connection. Existing pieces near the ladder's projected path often provide a convenient means of escape.

Figure 10.28 illustrates a ladder in progress along row 3. Black can't afford to keep playing along this line; if they continue with *28* E3, then White can play at point *p* or *q* to shut out the ladder and complete a solid connection that threatens to win the game. Black must attempt to engineer an escape as soon as possible.

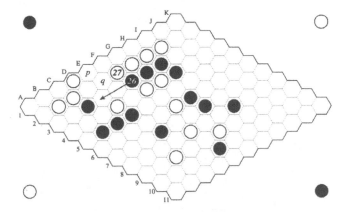

Figure 10.28. A ladder in progress.

Black's best play in this situation is *28* E2 (Figure 10.29). The path analysis shown in Figure 10.30 reveals this to be a forking ladder escape that cannot be beaten. The existing piece *2*, although not a recognized ladder escape template, provides an alternative disjoint path through chain (*6*, *12*, *18*) to piece *26*, which makes the escape successful.

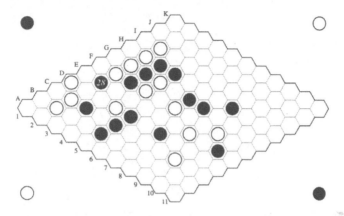

Figure 10.29. Black engineers an escape.

Black has now won the game, having a safe connection from the top left edge through to piece *14*. Piece *14* can in turn force a ladder along row 10, for which existing piece *10* provides the escape and the win (Figure 10.30).

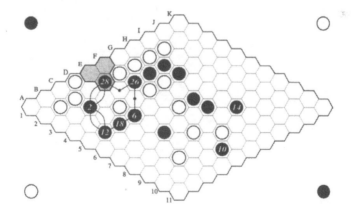

Figure 10.30. An alternative path guarantees the escape's success.

10.3.4 Confuse the Opponent

Sometimes the situation is so bad that defeat seems inevitable. However, if the opponent's winning connection is not trivial there's a chance that they can be confused with an apparently inconsequential side play that soon proves useful. If the game is already lost, it can't hurt to try such a *red herring* move. Consider the situation shown in Figure 10.31. White has just played **28** C2 to set up a very strong position.

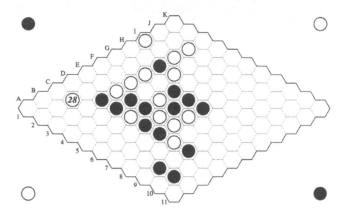

Figure 10.31. White's move **28** C2 puts them in a commanding position.

As shown in Figure 10.32, move **28** C2 provides White with both an escape piece for a ladder originating at piece **14** and an easy connection to the top right edge via piece **26**. It appears that White will win very shortly. Where should Black move to best advantage in this desperate situation?

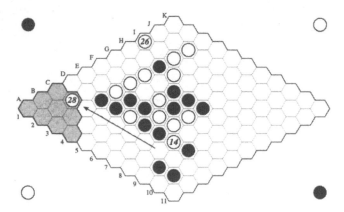

Figure 10.32. White has both a ladder escape from piece **14** and an easy connection to piece **26**.

Black chooses to play **29** A3 as shown in Figure 10.33. This move intrudes into the ladder escape template, but is little more than a red herring as White can easily negate this threat if they so choose and reconnect the escape template.

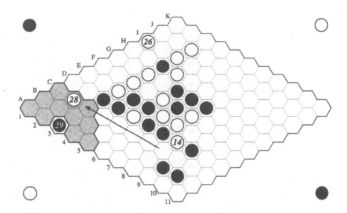

Figure 10.33. Black's template intrusion *29* A3 changes the nature of the ladder escape.

However, there's the chance that White will only observe that their escape formation has changed from edge template IIIb to edge template IIIc, and give it no more thought. This would be a fatal mistake, as the potential ladder path now interferes with the escape template, making it invalid.

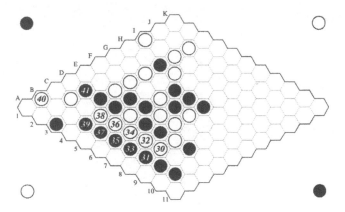

Figure 10.34. White's ladder fails to escape.

Figure 10.34 shows the result if White ignores the intrusion and instead chooses to force the ladder with moves *30-38*. Black's red herring piece at A3 threatens a crucial link to the top left edge following move *39* B4, which White must block with *40* B1.This leaves Black to play the killer move *41* D2.

Black has now won the game, as shown in Figure 10.35. Piece *41* provides an escape for a ladder originating at piece *25*, which is only one step away from connecting to piece *7*. White cannot stop *7* connecting to the lower right edge. White is forced to play *42* C2 at point *p*, to which Black can respond *43* G4 at *q* to put the win beyond doubt.

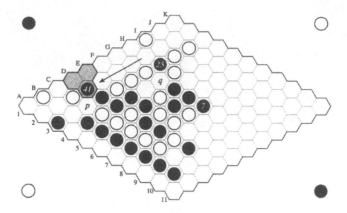

Figure 10.35. Black now has the winning connection.

White gave this game away. Black's red herring intrusion *29* A3 is trivial to defend against with a move such as *30* B4, as shown in Figure 10.36. This restores the ladder escape and maintains White's connection to the top along row 1.

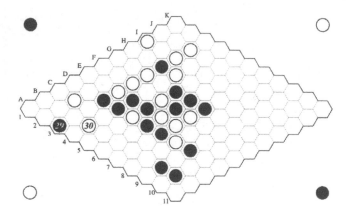

Figure 10.36. A simple defense of Black's intrusion.

The key points of this example are:

• *the danger of ignoring the opponent's last move (see Section 3.2.3), and*

• *confusing the opponent with a red herring move before resigning.*

As explained in Section 3.3.3, the exact point at which the game ends depends largely on the skill level of the two players. An experienced opponent will rarely be distracted successfully by a red herring move.

10.3.5 Extrapolate Ladder Connections

David Boll [1994] suggests that the player should treat ladders as a single unit to simplify board analysis. This suggestion can be extended to include treating:

- *the ladder,*

- *the group from which it originated, and*

- *the escape template*

as a single group connected to the edge. This is demonstrated in the board shown in Figure 10.37, where White has just played potential ladder escape *10* B10.

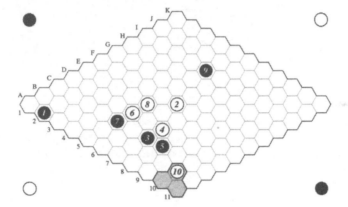

Figure 10.37. White's *10* B10 provides a potential ladder escape.

Black unwisely ignores the impending ladder through C6 and instead plays *11* D9 to block White's connection across the bottom of the board from piece *10* (Figure 10.38). White then plays *12* C6 to force the ladder.

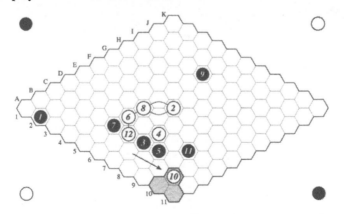

Figure 10.38. Black's *11* D9 allows White to force the ladder with *12* C6.

White's ladder originates from the group composed of pieces *2, 4, 6, 8,* and *12* is safely connected to the edge as shown in Figure 10.39.

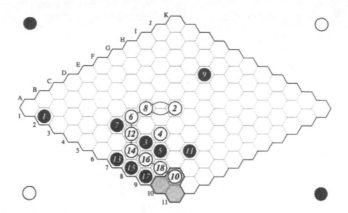

Figure 10.39. White's ladder is safely connected to the edge.

Piece *12* (and by connection pieces *2, 4, 6,* and *8*) can be considered to be connected to the lower right edge. The area required for this connection is *the maximum area required for the opponent's best defense* as indicated by the shaded region in Figure 10.40. This forms a group consisting of:

- *the lower left edge,*

- *the pieces 2, 6, 8, 10, and 12, and*

- *the empty points indicated.*

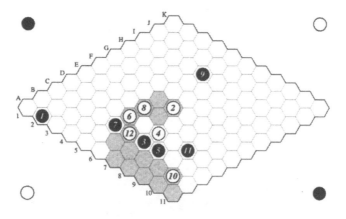

Figure 10.40. Minimum area required for the ladder escape connection.

The player should learn to view such safely connected groups as single units from which to build further connections. This process is described in detail in Chapter 8, Algorithmic Board Evaluation.

10.3.6 Make Escape Pieces as Threatening as Possible

When a ladder forms but no forking escape presents itself, the player's best strategy is to play a potential escape piece that is as threatening as possible, in the hope that the opponent will choose to block the threat and leave the ladder escape unhindered.

Figure 10.41 shows a case in point. White is in a strong position, and with *12* I7 has pushed through for a ladder that Black is obliged to block with *13* K6.

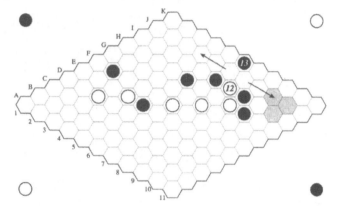

Figure 10.41. White to move and make a ladder escape.

White's practical options for playing an escape to the right of the ladder formation are limited to the region shaded. These moves are not very threatening in themselves and Black would be safe blocking the ladder at J7 on their next turn.

Let's now look at escape moves to the left of the ladder formation.

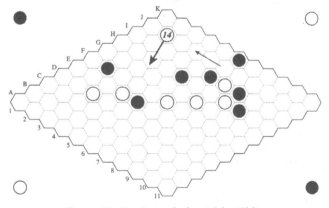

Figure 10.42. A good play *14* by White.

Move *14* J2 shown in Figure 10.42 is a good escape move by White. If Black blocks the ladder, then *14* J2 threatens to form a solid connection across the board following a move at H3. Even though this is not a forking ladder escape as such, Black would be wise to block this connection and concede the ladder to White. This is a good example of a move serving two purposes.

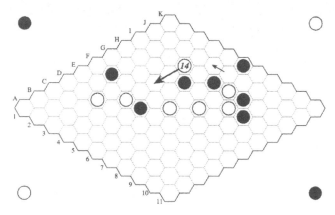

Figure 10.43. Another good play *14* by White.

Another good play *14* I4 by White is shown in Figure 10.43. This move also threatens to push across the board if not answered, but is not a direct forking ladder escape. *14* I3 would be a bad play as Black could then foil the escape by moving at I4 themselves.

10.4 Looking Ahead

Hex is first and foremost a combinatorial game; the best way to evaluate any given board situation is to play it through as thoroughly as possible. The player may be obliged to move within certain time constraints, so it's not practical to consider the consequences of every move. Instead a few key moves should be identified and played through deeply to determine whether their outcome is favorable.

If the game is being played remotely over several days (or even weeks or months) then each player has the luxury of working through moves with pen and paper (or their modern equivalent, the text editor) which allows a more thorough and often complete analysis of the game. The blank Hex boards provided in Appendix F can be copied for rough working.

However, if the game is being played over-the-board, promising lines must be played through mentally which can become extremely difficult after only a few moves. Players tend to narrow down the search to as few choices as possible each turn and may therefore overlook some good moves. Given a limited amount of time, the player must balance exploring a few key moves deeply with exploring a large number of moves shallowly.

Templates save time and effort by allowing the player to evaluate the connection within a specific region of the board at a glance. Other situations crucial to the player's connection and not covered by known patterns just have to be played through using lookahead search. A distinction is made between *short* and *long* range search:

•   *short range search is exhaustive and explores a limited situation thoroughly, and*

•   *long range search explores a few avenues deeply and requires that some (possibly unsound) assumptions be made along the way.*

10.4.1 Short Term

Short term lookahead resolves key plays that require only a few moves for completion. This gives a very accurate or often complete analysis of the situation. For example, consider the board illustrated in Figure 10.44 with White to move. The connections of both players are well defined except for the region in the bottom corner of the board. Within this region points *p* and *q* look like promising moves, but it's not obvious which is better. White applies a short term lookahead to determine their outcome.

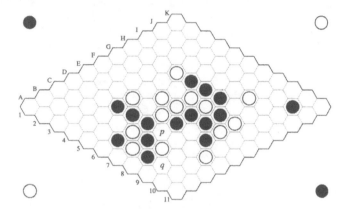

Figure 10.44. Moves *p* and *q* look promising for White.

Figure 10.45 shows the sequence of play that White imagines will occur following a move at point *p*. It should be assumed that Black makes the optimal move at each turn, to anticipate the worst case scenario.

If White chooses to play *29* D7 at *p* then Black's best reply is forcing move *30* B9 at *q*. It appears that *q* is indeed an important point. White is then forced to reply *31* B10 and *33* C10, giving Black the win with *34* E10. Black's winning connection is shown in Figure 10.46.

Figure 10.47 shows the projected result if White instead chooses to play *29* B9 at point *q*. This is a very threatening forcing move, and White is then able to play a further sequence of forcing moves *31*, *33*, and *35* leading to a winning connection (shown in Figure 10.48).

Move *q* is the correct choice in this case, and is the difference between victory and defeat. The usefulness of short term lookahead was demonstrated in Section 10.1, where closer analysis of three similar moves revealed only one of them to be a winning move.

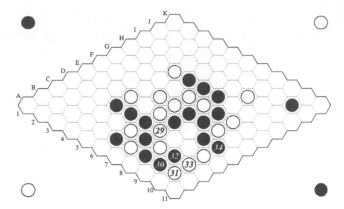

Figure 10.45. White does a short lookahead on move *p*... and sees a loss.

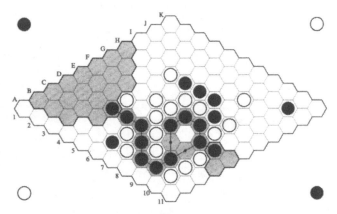

Figure 10.46. Black's winning connection.

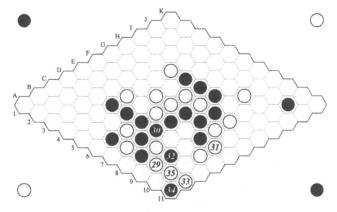

Figure 10.47. White does a short lookahead on move *q*... and sees a win.

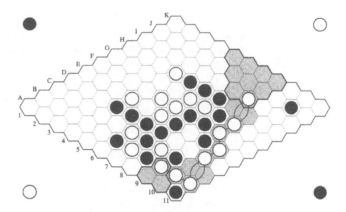

Figure 10.48. White's winning connection.

10.4.2 Long Term

In contrast to short term lookahead, *long term lookahead* explores fewer probable lines of play to greater depth (less breadth of search). The result is generally less accurate than the short term search unless composed primarily of forcing moves that limit the opponent's replies.

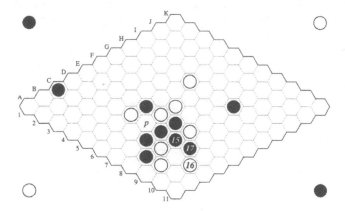

Figure 10.49. White to move. Black is one step from a win through point *p*.

Long term lookahead is used primarily for gauging how a position is *likely* to develop, allowing the player to prepare for the anticipated result. Figure 10.49 illustrates a situation where White is in trouble and performs a long term lookahead to search for a possible way out. This example is drawn from the sample game annotated in Section 10.2.

Black's pieces *15* and *17* force a ladder along row 9 as indicated in Figure 10.50. This ladder is not guaranteed to escape successfully as piece *5* at H8 does not constitute an escape template for 3-row ladders (see Figure 7.12 for a play that defeats it), but this position looks extremely dangerous for White nonetheless.

Black is now in a strong position as piece *9* is safely connected to the top left edge via escape piece *1* (the reader may like to prove this for themselves). Notice how the seemingly inconsequential opening move *1* C1 has a major impact on the game.

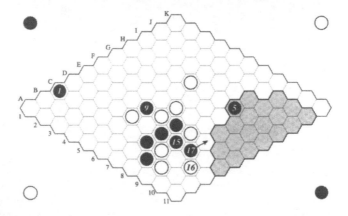

Figure 10.50. Black's ladder from piece *17* threatens to connect via piece *5*.

It appears that White's only sensible reply is the forced move *p*. However, they decide to first perform a long term lookahead to see if any other opportunities are likely to develop from this desperate situation.

White starts their lookahead with the obvious move *18* D6, to which Black's best reply is most likely *19* C5. This allows White to play the excellent move *20* C3 that threatens a ladder escape and also provides an avenue of connection along the top of the board. Black foils the escape, and White is able to push across the board with forcing moves *22, 24, 26,* and *28*.

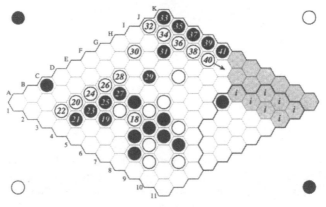

Figure 10.51. Long term lookahead reveals a probable scenario.

White then anticipates forcing their way towards the top right corner with moves *30* and *32*, then forcing a ladder down column J as indicated in Figure 10.51. This ladder is not to White's advantage unless they can engineer an escape within the shaded region of the board. Possible escape points that also intrude on the template of Black's escape piece *5* are marked *i*. These are the most desirable points to occupy.

Assuming that this sequence is the most likely to develop, White's problem then becomes one of organizing an escape piece in the shaded area. They cannot simply intrude into the template by playing at one of the points marked *i* as the foiling move would occur on the wrong side of the ladder escape for a ladder progressing down the board (see Section 7.5.2).

White sees a way of achieving this goal by playing out Black's ladder along the bottom of the board, then pushing past the potential escape piece when it is reached. This play allows White to steal some territory in the desired area, and cannot harm their position as Black's ladder is all but safely connected.

Figure 10.52 shows the result of White forcing Black to play the ladder to its conclusion with *25* G8. White then plays the key move *26* H9, a forcing move that intrudes on Black's safe connection and steals some territory. Black's reply *27* I8 is a good defensive move. It both restores the ladder connection and foils *26* as an escape piece. White might now be tempted to play at point *q* to attempt another ladder escape, but Black can defeat this play as it stands. White's only hope is to establish the ladder down column J as foreseen in the lookahead, then to play at *q* just before the ladder reaches it.

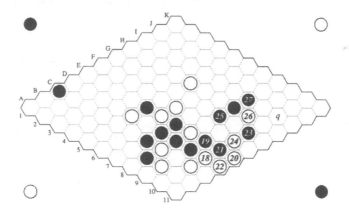

Figure 10.52. White plays through the ladder.

White's best plan is therefore to proceed with play as predicted by this long term lookahead and form a ladder down column J.

Things do not work out as expected for White however, as shown in Figure 10.53. Black plays the killer move *39* G3 to win the game prematurely. Although sound in principle, White's lookahead failed to consider this move and assumed that Black would play the

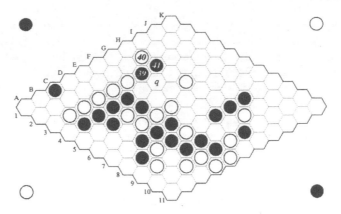

Figure 10.53. How the game actually unfolds.

more obvious move at point *q*. This single bad assumption in White's chain of reasoning was fatal, and it's now clear that lookahead move *29* (actual move *39*) was a pivotal play, and worthy of additional lookahead in its own right.

In this case lookahead has been used not so much to choose between otherwise equally good moves as with the short term example, but to gauge how the game might develop and plan accordingly. Note that long term lookahead is useful in developing overall strategies, but can be dangerous if relied upon as a faithful prediction of future play; it should only be considered an *estimation of the most likely development of play*.

Towards the end of a game, when both players' connections are well-defined and forcing moves cannot be ignored, short term lookahead should be used profusely. Lookahead is less accurate earlier in the game, as forcing moves are not so pressing and the opponent may choose to ignore them and play elsewhere, invalidating the line of play being investigated.

Long term lookahead should not be used until both players have at least a few pieces on the board. There are just too many possibilities to consider to any depth, and those paths followed will most likely prove to be wildly inaccurate. It's surprising how quickly two games can diverge even from identical opening sequences.

In a match between two equally skilled opponents, the game is usually won by the player who is prepared to look ahead the furthest. This can be extremely hard work and require a lot of assumptions to be made regarding the opponent's moves, but it pays off.

10.5 Overall Game Plan

Now that the major points of Hex strategy have been covered in some detail, it's possible to describe an overall game plan. This will vary depending on the current *state* of the game: whether the game is won or lost, or to whose advantage it is currently poised.

If the game is in its opening stage then the player should apply the opening strategy discussed in Chapter 9. This involves selecting an appropriate opening move based on the opponent's skill level, deciding whether to swap, and how to best defend against a move in the central region.

If the game has advanced beyond the opening stage the next step is to apply the algorithmic board evaluation described in Chapter 8 to determine the best spanning path for each player. Let's call the best spanning path of the player with the move in hand $best_p$ and the opponent's best spanning path $best_o$.

Comparison between $best_p$ and $best_o$ indicates the current board state as described below. The player's best approach for each state is described. Note that it's not possible for both $best_p$ and $best_p$ to equal zero (see Appendix D.5). The expression $best_p > best_o$ indicates that the player's best spanning path has a better connectivity (its connectivity value is lower) than the opponent's.

10.5.1 Game Won

Condition: $best_p = 0$

The player has a safely connected spanning path and cannot be beaten. The quickest way to finish the game is to fill in the pivot points within the winning path. Any intrusions into the winning path by the opponent should be answered immediately by playing in their dual point.

10.5.2 Game Lost

Condition: $best_o = 0$

The opponent has a safely connected spanning path and should win the game unless they make a serious error. The player may attempt to confuse the opponent with a couple of red herring moves that intrude into the opponent's winning path. If this fails and it's obvious that the opponent can secure their connection, then the player should resign immediately and get on with the next game.

10.5.3 Player's Advantage

Condition: $best_p \geq best_o$

This condition includes the case $best_p = best_o$ as the player has a move in hand and will (hopefully) improve their connection this turn.

The player's first priority in this situation is to play defensively and develop their connection. Escapes should be prepared for any ladders likely to form. This is best done with forcing moves or forks that the opponent is forced to counter immediately.

10.5.4 Opponent's Advantage

Condition: $best_p < best_o$

The player is in a losing situation and must try to impede the opponent's connection wherever possible. They should play forcing moves in weak links along the opponent's best spanning path, and not allow them to make any free moves or to dictate play. The situation is desperate and calls for aggressive moves intended to unbalance or confuse the opponent.

The player should identify the weakest links in their and their opponent's best spanning paths. If these weak regions overlap, play there. A simultaneous improvement in the player's connectivity and decrease in the opponent's connectivity may swing the momentum of the game.

If really stuck, the player should just move where they believe the opponent would most like to play next turn.

10.5.5 General

In general, the player should:

*   *foil any potential ladder escapes belonging to the opponent (or at least guarantee that they can be foiled at will with a future move),*

*   *look for possible ladder formations and work to engineer escapes for themselves,*

*   *always be on the lookout for forking moves and forked ladder escapes,*

*   *play forcing moves that gain territory where possible, and*

*   *play in the weakest link of their best connection, the weakest link of their opponent's best connection, or both if possible.*

Summary

The opening moves set the course for the entire game. The player should try to gauge the opponent's level of skill and play the strongest opening that will not be swapped. If the second move is made in the central hexagon they should defend against it immediately.

Forcing moves should be recognized as either benign or dangerous. The player should try to limit the number of dangerous forcing moves they leave on the board.

Foil potential ladder escapes if possible. This may be delayed to distract the opponent, but must be attempted before the ladder forms.

Look ahead to determine the repercussions of possible moves. This is useful for choosing the optimal move from a set of likely candidates, or for forecasting the general progression of the game in order to prepare traps such as ladder escapes. In a match between two equally skilled opponents, the game is usually won by the player who is prepared to work harder and perform the more comprehensive lookahead.

The player should work to an overall game strategy. This should be flexible to accommodate novel board situations and different states of play.

Annotated Sample Games

The following sample games illustrate points of strategy discussed above in the context of actual game play. All examples shown are drawn from games played on the Gamerz.NET server [Rognlie 1996].

11.1 Notation

The following notation is used to describe moves:

31 H5   Thirty-first move of the game, played at point H5.

31 H5!   A good move.

31 H5!!   An excellent move (swings momentum in the player's favor or sets up a win).

31 H5?   A questionable move.

31 H5??   A disastrously bad move (potential game loser).

1 A1 Swap   The second player has elected to play the swap option. They take the opponent's color and effectively steal the first move.

More elaborate systems for annotating moves in the context of a game exist, but the system outlined here is more than adequate to highlight the key plays in a game. For instance, some systems include notation such as *12* G2?! to indicate an apparently bad move that later proves to be useful. This is not particularly relevant in a purely deterministic game such as Hex; if a piece later proves useful, then it's a good move. Luck may play a part if a player makes a random move beyond their understanding of the game that later proves useful, but the annotator cannot presume to guess the players' motives to this extent.

Although it can never harm a player's position to have one more piece on the board, a player can still make bad or disastrous moves. Such wasted moves may allow the opponent to complete an attack that should have been blocked, or let the momentum of the game swing the opponent's way. To have a move in hand is a huge (often winning) advantage in Hex, particularly for close games between experienced players.

11.2   Ladder Escapes Denied

The following game was played between two experienced players, and provides a number of interesting passages of play that demonstrate points of strategy in action. Notice in particular Black's expert foiling of ladder escapes.

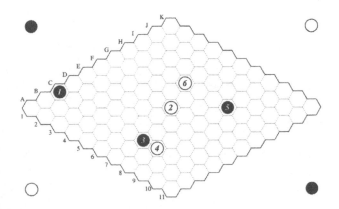

Opening Moves

Black   White

1 C1   A very cautious opening.

         2 F6   White declines the swap and takes the commanding center point.

3 C7   This move is the traditional defense against F6.

         4 C8!   A good reply which puts *3* immediately under pressure, threatens to connect *2* with the edge, and provides a potential ladder escape for a 2-row ladder down column B.

Early Game

Black   White

5 H8   This piece is safely connected to the edge by a 4-row edge template, and potential ladder escape. Black is preparing well ahead.

         6 H5   White strengthens the weakest link along their best connection. This move also reduces the threat of a connection from *5* pushing up through this region.

White is now well placed, and is already one move away (D7 or E7) from a winning connection. These are the vulnerable points of the loose connection between pieces *2* and *4*.

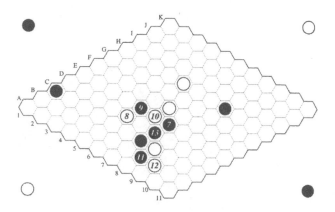

Black   **White**

7 E7?
          Black obstructs the connection between *2* and *4*, but this is a bad choice[1].

          8 D5?   White fails to notice winning moves E5 and E4 (see footnote).

9 E5
          Intrusion in one of the vulnerable points between *2* and *8*.

          10 E6!   White makes the most of a forced reply by forking potential connections to both *4* and *8*. If Black blocks the obvious connection with D7, White can force a ladder from piece *8* and win the game.

11 B8!!
          A great ladder escape foil held off until it was needed.

          12 B9   White completes the connection.

13 D7
          Black can now safely block the obvious connection.

By delaying the necessary blocking move *13* by one turn in order to play the ladder escape foil *11*, Black has swung the game in their favor. Not only did Black avoid imminent

---

[1] Leonid Gluhovsky points out that *7* E7 is a potentially losing move for Black. Luckily for them White fails to see the winning reply *8* E5. If *9* D5 then *10* D6, *11* A8, *12* B6, *13* D6, *14* D6, *10* D6; if *9* B7 then *10* D6, *11* D7, *12* C5; if *9* B6 then *10* D6, *11* B8, *12* C6, *13* B7, *14* C4. Black should have played *7* D7. *8* E4 is also a winning move for White.

defeat, but they were able to force a connection between pieces *3* and *7* which opens up a passage to both edges.

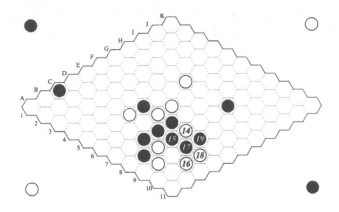

Middle Game

Black   **White**

14 E8   White attempts to impede Black's progress and creates a bottleneck formation.

15 D8   Black pushes through...

16 C10   ... and White plugs the gap.

17 D9   Black forces a 3-ladder.

18 D10   White blocks.

19 E9   Black responds and pushes the ladder along.

   Black's ladder along row 9 has an easy escape with piece *5* and is guaranteed to connect with the lower right edge. So why is White continuing to force the ladder along? They are making the most of a bad situation and maneuvering play towards the rightmost corner of the board in the hope that they will eventually be able to play a helpful forcing move, such as a potential escape for ladders coming down rows I or J. This plan shows a good deal of forward planning on White's behalf.

   This ladder now becomes very important to Black. As long as White's moves intrude on this ladder's connection, they can continue this play safe in the knowledge that Black is more or less forced to reply rather than move elsewhere.

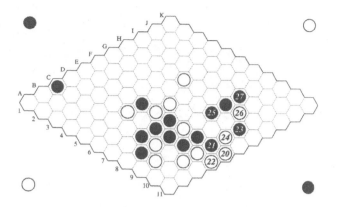

Black   White

20 E11   White yields a row.

21 E10   Forcing move that pushes the ladder one row closer to the edge.

22 D11   Forced reply.

23 G10   Another escape that guarantees the ladder's connection.

24 F10   White persists with their attack.

25 G8   The connection is now safely completed.

26 H9!   White finally achieves what they set out to accomplish. This piece is
            safely connected to the top right edge, and provides a potential escape
            for ladders down columns I and J.

27 I8!   Once again Black plays an excellent ladder escape foil, rather than the
            more obvious G9. The proximity of piece *5* makes this foil successful.

White now has the opportunity of playing I10, a move that threatens both a winning
connection across the bottom and provides an escape for a ladder down column J. This
fork looks dangerous but can be foiled by Black at vulnerable point J9. However, White has
a slim hope that their opponent may misplay the foil, and it is now in their best interest to
force a ladder down row J.

White does not immediately play at I10 but keeps this move up their sleeve, in the hope
that Black does not notice it and can be maneuvered into forming the necessary ladder.

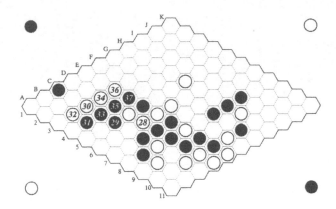

Black	White	
	28 D6	Play returns to the middle, where White attempts to push through to the left.
29 C5		Black responds by blocking with a bottleneck formation.
	30 C3!	Attempted ladder escape fork. This piece is one space removed from the nearest enemy piece along the ladder line (piece 29), so Black must be very careful here.
31 B4		Ladder escape foil.
	32 B3	White completes the connection to the edge.
33 C4!		Black cuts off this ladder escape.

Black has a solid connection across the region between pieces (3, 11) and (29, 31, 33). White must now push across the top from piece 30 and hope to force a ladder down column J.

End Game

Black	White	
	34 D3	Forcing move.
35 D4		Forced reply.
	36 E3	Forcing move.
37 E4		Forced reply.

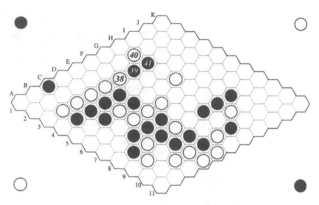

Black   White

38 F3   White appears to be in a strong position and threatens to connect with pieces *2* or *6* for a win. If Black attempts to block this connection with the obvious G4, then White is set to force a ladder down column J, during which they can play the killer move I10 to win the game.

39 G3!!   A brilliant move which reduces all of White's options to losing lines of play.

Closing Moves

Black   White

40 H2   Last ditch attempt by Black to force a ladder along row 2 which will then result in a ladder down column J.

41 H3!   Another very strong move that averts any possibility of defeat. The chain (*39*, *40*) is safely connected to the edge and threatens forking moves at F4 and G5, each of which result in a winning connection for White.

This game demonstrates several of the strategic points previously discussed, in particular laddering. Both players showed good use of ladders and ladder escapes, and White played some brilliant ladder escape foils and forking moves to win the game. Also note White's foreplanning in preparing a ladder escape with move *26* that would not be required until approximately move *50*. This demonstrates just how far a player can plan ahead if they concentrate on very narrow avenues of attack.

11.3   Premature Resignation

This game was played between an expert (Black) and an intermediate player (White). It is notable due to the fact that White outplays their more experienced opponent and has at least two winning plays, but fails to recognize these and eventually resigns the game! This highlights the fact that a player should not resign until the outcome of the game is beyond doubt, and it is clear that both players realize this.

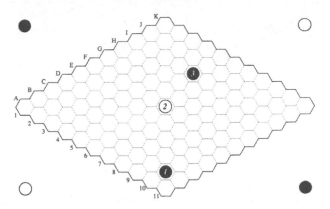

Opening Moves

Black   **White**

1 B10   Opening on the short diagonal, but safely away from the central point F6 in case White decides to swap. This is a deceptively strong move, and Black probably hopes that White does not swap.

      2 F6   White declines the swap, but plays the solid center point F6.

3 I5   Black starts to Block White's *2* from the top right edge well away from it. This is the traditional defense against an opening at F6.

    Many games of Hex start with this sequence or one very similar.

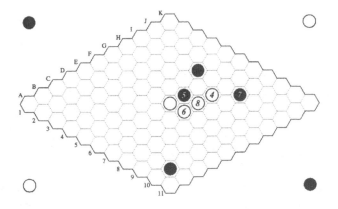

Early Game

Black **White**

4 H7   This piece is well placed to connect to both *2* and the top right edge. It is a threatening move that Black can not ignore.

5 G6   Black plays in one of the vulnerable points of the loose connection between pieces *2* and *4*. This is an aggressive move that puts pressure on piece *2* and threatens to connect to the bottom right edge.

     6 F7

7 I8!   Fork: *7* is a potential ladder escape, and also threatens to connect with Black's *3*, requiring a reply.

     8 G7?   White plugs the gap and avoids *5* making a ladder to *7*, but instead should have moved to foil the fork at *7* (a move at H8 would have achieved this and been much better in general).

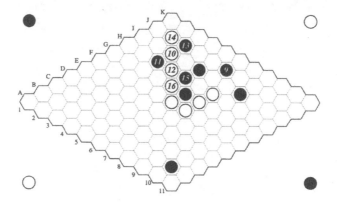

9 J6   Black completes a formidable connection.

     10 I3

11 H3

     12 H4

13 J3?   An inconsequential move, maybe J2 would have been better.

Black   White

14 J2

15 H5??   This is a terrible move. It threatens to make no new connections and
                   gains no territory, and is a wasted move. White now has the momentum
                   and can continue to attack and dictate play.

16 G5

White is now in an excellent position and has split Black's defenses, which are in
tatters. White's main body of pieces has a safe connection to the top right edge and is
threatening to push across the board. Things are looking very bad for Black, whose poor
play has led to this situation—perhaps overconfident and not playing carefully enough
against a less experienced opponent?

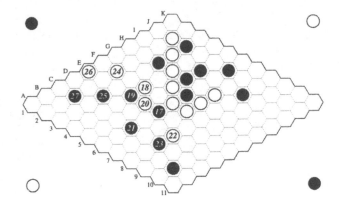

Middle Game

Black   White

17 E6   Black tries desperately to impede White's progress across the board
                   with an aggressive, close-in move.

18 F4   White is able to step around 17 without any trouble.

19 E4   Ladder formation: Black forms a bottleneck.

20 E5   White pushes through.

21 C6   Black plugs the gap.

22 D8   White attempts to push through around the side.

Black   White

23 C8?   Black blocks easily, but leaves a possible forcing move at B9 that could
                be used to the opponent's advantage. A9 might have been a better play.

         24 F2

25 D3

         26 E1

27 C2

Black has regrouped due to White's indecisive play in the last few moves, and appears
to be have recovered the game somewhat. However, White has a reasonably clear win if
they play well.

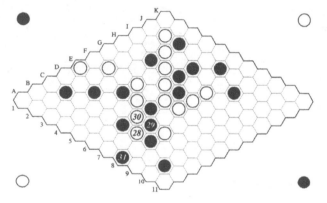

End Game
Black   White

         28 C7!   A very threatening move, *28* is safely connected to the top right edge
                    via path consolidation to pieces at *20* E5 and *22* D8, and is very close to
                    connecting to the bottom left edge.

29 D7   Forced reply.

         30 D6   White realizes the consolidation, and is very close to victory.

31 A8   Forced reply to avert immediate defeat.

White now has a winning play at B9. Black has not resigned yet despite imminent defeat, and waits to see whether White has noticed the winning play. If White plays B9 the game is effectively over, and any further closing moves redundant.

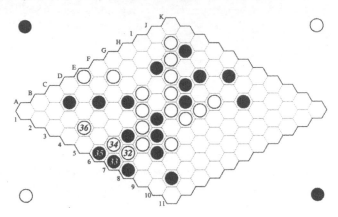

Closing Moves

Black   White

	32 B7?	White has apparently failed to notice the winning move at B9 and tries to connect elsewhere.
33 A7		Forced reply.
	34 B6	White forces a ladder.
35 A6		Forced reply.
	36 B4	A feeble attempt at a ladder escape (no fork), which proves to be a useful move shortly.

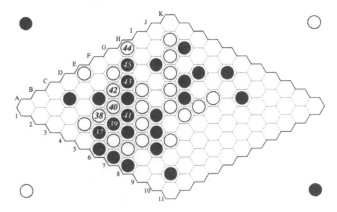

Black   White

37 B5   Forced reply.

       38 C4

39 C5   Forced reply.

       40 D4   White threatens to connect to the main group.

41 D5   Forced reply.

       42 E3   White has barged through the defense and is looking dangerous.

43 F3   Forced reply.

       44 H1?? G2 would have completed White's connection across the top with a
       forcing move and secured a win (again).

45 G2!   Black takes the initiative by making a forcing move of their own, severing
                  White's connection across the top. If White replies with move G1, then
                  Black counters with I2 and the connection is still lost.

    46 Resign?? Frustrated by the broken connection across the top, White resigns with-
    out noticing the winning move at B9.

Black survives! This victory is remarkable due to White's failure to notice a winning
move for the last third of the game, and Black's ability to swing the momentum of play with
a single move (move *45*) following a run of eight consecutive forced replies.

11.4 How Not to Play

The following game demonstrates just how short a game can be if the opponents are
mismatched in level of skill. Black is a beginner with little knowledge of Hex strategy, and
White is an intermediate player with knowledge of bridges and edge templates. This is all
that is required to achieve an easy victory.

Opening Moves

Black   White

1 A4?   A weak opening which Black will regret if their opponent does not swap.

      2 F6   White sensibly declines the swap.

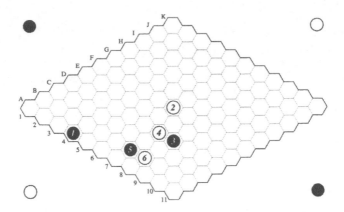

Black   White

3 D8??   A wasted move that has no effect on the game. It is does not block the strong *2* F6 and does not form an edge template to the edge. C7, the traditional reply to F6, would have been far preferable. Black has lost the momentum already.

Early Game
Black   White

         4 D7   Solid bridge move from *2* that forms a 4-row edge template.

5 B7?   Black fails to realize that *2* and *4* are now safely connected to the lower left edge and intrudes into the template.

         6 B8   White completes the connection to the edge.

We have reached the middle game already, and Black is in big trouble. Their best spanning path is an extremely weak one, and their defense is in tatters. To give an indication of the disparity in the positions, White's best spanning path is 1-connected, while Black's best spanning path is 3-connected.

Black must now attempt to defend on the right hand side of the board.

Middle Game
Black   White

7 H5   A passable blocking move, but I5 would have been better.

         8 G7   White extends their main body of pieces with a bridge and looks very threatening.

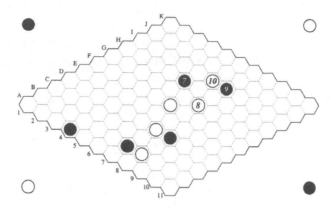

Black **White**

9 I7?? Black's best chance here would have been to attempt to force a ladder, perhaps by playing at H7. *9* does not even block White's progress to the edge.

10 I6 An easy win.

This game is a lesson in how *not* to play Hex. Black did nothing more clever than build bridges from their central group; it was won solely through White's poor defense.

11.5 Another Point of View

The following is an abbreviated version of an annotation provided by Tom Hayes, giving another player's view of how a Hex game may be interpreted. The game was played between two experienced players.

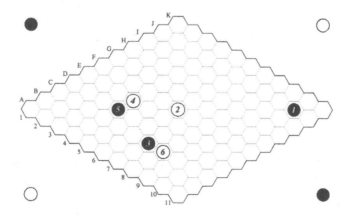

Opening Moves

Black   White

1 J10   A very strong opening for Black, to the extent that it should not be played against a player who will swap.

         2 F6   The most natural move, but swapping is probably superior in this case.

3 C7   Standard defensive move against F6.

Early Game

Black   White

         4 E4   D4 is the usual choice here; however, E4 works out very well in this game.

5 D4?   Seems like excellent defense at first—it connects to the top, forcing White to approach the side via D5 if they wish to hook up E4 directly. However, it is more or less refuted by the game line. *5* E6 appears to be a promising in-between or *zwischenzug* move, threatening to connect to the top at F4, and taking away a useful space from White.

         6 C8   A move which works directly towards White's primary goal at this stage— connecting F6 to the left side. Now it is too late for Black to play E6.

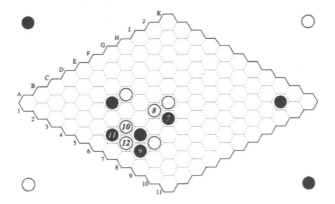

7 E7   At this point Black should probably have switched tacks and played *7* I5. Things look pretty bad on the left side as we'll see by move *10*. Still, Black may be able to profit by making the right attacking move on the left later. Which move this is may depend on developments on the right side.

Black   White

8 E6!   White, on the other hand, must continue responding to Black's threats on the left side. This move is a bit of a zinger, threatening both *10* D7, connecting to the left, and *10* C6 which is almost as good.

9 B8   Challenges the connection of C8 to the edge, while preventing any ladders down to it along the B column, hence is often a strong move. In this case, however, it is too late.

10 C6   Seems to guarantee White a connection to the left edge at F6 or H4.

11 B6

12 B7   Two apparently inconsequential but obvious mistakes. White's correct response is *12* C5, which dominates *12* B7. In light of this response, we see Black's move *11* B6 as worse than useless.

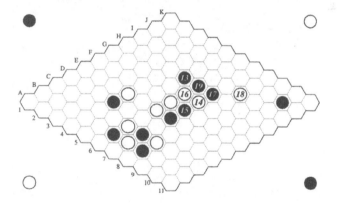

Middle Game

Black   White

13 H5   Presumably inspired by their pieces on rows 7 and below, Black plays aggressively on the right side.

14 G7   The direct approach.

15 F7

16 G6   Might as well.

Black White

17 H7

        18 I8

19 H6??   Here Black makes a terrible blunder. *19* I5 would have given them a
                 winning position. As it is, White can win with such moves as *20* B9, *20*
                 E5 or *20* G9. Why is this a terrible blunder? *19* I5 would have given Black
                 a second piece on row 5 for use in connecting to the top. Additionally,
                 White can no longer ladder down column I. White's forcing move I6 is
                 useless to them as they already have the piece on I8.

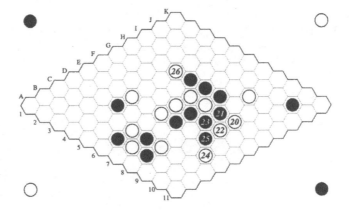

End Game

White maintains their winning connection throughout the remaining moves. The board
position at this point is discussed in detail in Section 6.1.1.

Black   White

        20 G9   Move *20* B9!! would have made White's win obvious at this point. If
                 Black plays *21* H4 in response to *20* B9, then White wins with *22* G9, *23*
                 G8, *24* F9, *25* F8, and *26* D10. If Black plays *21* G9 in response to *20* B9,
                 then White wins with *22* H4, *23* G5, *24* G4, *25* F5, *26* F4, and *29* D5 or E5.

21 G8
        22 F9
23 F8
        24 D10
25 E9
        26 H4

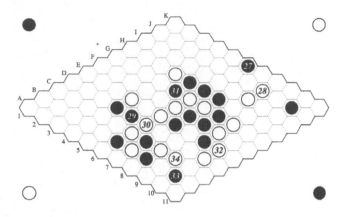

Closing Moves

Black	White
27 K6	
	28 J8
29 D5	
	30 D6
31 G5	
	32 E10
33 B10	
	34 C9
35 Resign	

12

Strategy IV: Essential

This chapter describes the most important aspects of Hex strategy in a nutshell. It is intended to give the reader a quick introduction to the game in order to facilitate good play as soon as possible. It may take a few games for the relevance of these ideas to crystallize, but after reading and understanding this material the reader should be able to play at a reasonable level.

These key points are drawn from previous chapters and presented here in summarized form to provide a concise Hex strategy reference. Sections that contain more in-depth discussion of each point are indicated and should be consulted if the player is to develop their game further. The player should also familiarize themselves with the additional points of strategy described in preceding chapters to develop a well-rounded game.

12.1 Opening Play

The central hexagon F6 is the strongest point on the board and the best choice for opening move. This gives the first player a significant advantage and almost guarantees a win unless they make a mistake. For this reason it is common to enforce the *swap option*:

• *The player to move second has the choice of swapping colors, effectively stealing the first player's move.*

This makes the first player think twice about opening at F6 and encourages them to play a more conservative opening move instead.

12.1.1 No Swap

If the swap option is not allowed then the first player should open at F6. The second player's best reply is to block this piece from one of the opponent's edges by playing at either G3 or E9, as shown by move *2* in Figure 12.1. [See Sections 9.1.1 and 9.1.2 for further details]

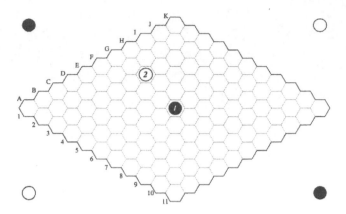

Figure 12.1. Opening move and reply (no swap option).

12.1.2 Using the Swap Option

If the swap option is allowed, then it's wise for the first player to choose a more conservative opening move. There is still considerable advantage in having an extra piece on the board, so they should endeavor to make the opening move strong enough to use in case they are left with it, but not so strong that the opponent wishes to swap it.

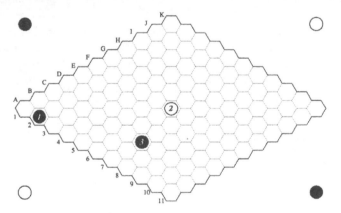

Figure 12.2. Opening play under threat of swap.

The opening move *1* A2 shown in Figure 12.2 is a good choice for opening move under threat of swap. It offers several benefits to the opening player but appears inconsequential enough that the opponent will usually decide against the swap and instead take the strong central hexagon as their second move, as is the case with White's reply *2* F6. Black's best play in this situation is *3* C7, which both blocks White's move *2* and threatens to connect with piece *1*. [See Sections 9.1.3 and 9.2]

12.2 Start Blocking at a Distance

In order to block an opponent's connection, it's tempting to play close to the leading piece of their attack. However, it is generally more effective to prepare blocking moves some distance from the advancing connection.

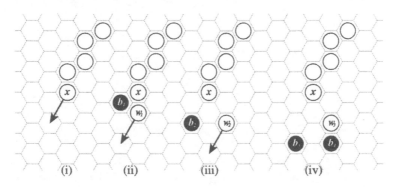

Figure 12.3. Three attempts to block White's advancing connection.

Figure 12.3 illustrates the importance of blocking at a distance. White has a chain of pieces safely connected to the top left edge and is threatening to advance towards the bottom right. If Black attempts to deflect White's advance piece x with the adjacent block b_1 then White can flow around unimpeded with move w_1. Even a close non-adjacent block such as b_2 is ineffective, as White can use bridge $w2$ to once again move around the block and continue the attack.

Reply b_3 to White's advancing piece x proves to be a more formidable block. White's attempt to step around the block with bridge w_3 can be successfully countered by move b_4.

The defensive block illustrated in Figure 12.3 (iv) is called the *classic defense*. [See Section 3.1.3]

12.3 Bridges

The configuration shown in Figure 12.4(i) is called the *bridge*. The two Black pieces are effectively connected, as possible White intrusions w_1 and w_2 can be easily countered

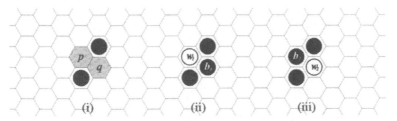

Figure 12.4. The bridge pattern and futile attempts to block it.

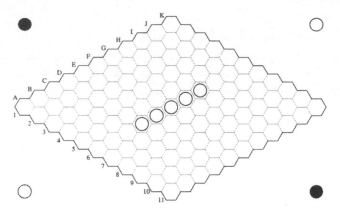

Figure 12.5. Four adjacent moves from the central hexagon.

with moves b_1 and b_2, respectively. The bridge connection is safe when considered in isolation, but can be defeated if the intruding move poses additional threats that must be answered first.

A bridge move covers twice the board distance as an adjacent move and allows the player to spread their connection across the board more quickly and with relative safety. Figures 12.5 and 12.6 show the comparison between a player choosing to spread their connection with adjacent moves to a player using bridge moves.

The four bridge moves shown in Figure 12.6 cover twice the distance that the adjacent moves cover. [See Sections 2.3.1, 3.1.1, and 5.1.1]

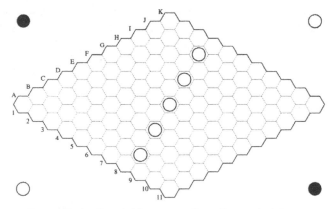

Figure 12.6. Four bridge moves from the central hexagon.

12.4 Play Defensively

One of the fundamental properties of Hex is that one player must win: a tie is not possible. Therefore, the player's task of completing their connection and the task of preventing their opponent's connection satisfy the same goal. Good defensive play will win a game as quickly as strong aggressive play, and in many ways *strong defense equals strong attack*.

It is a good idea to play defensively unless there is a good reason not to do so. If no obviously good move presents itself, a good rule of thumb is to *play where the opponent would most like to move next turn*. This strategy of foiling the opponent's next move is sound, as weakening the opponent's position will always improve the player's position.

Before making a move, players should always ask themselves "*what is the most damaging reply the opponent can make?*" [See Sections 3.3.1 and 3.3.2]

12.5 Edge Templates

A connection template is a predefined board pattern of a certain guaranteed connectivity. Generally we are more interested in *safe templates* that guarantee a connection even if the opponent has next move. There are two distinct forms of connection template:

- *interior templates* between pieces on the board (adjacent and bridge moves), and

- *edge templates* between a piece and its nearest home edge.

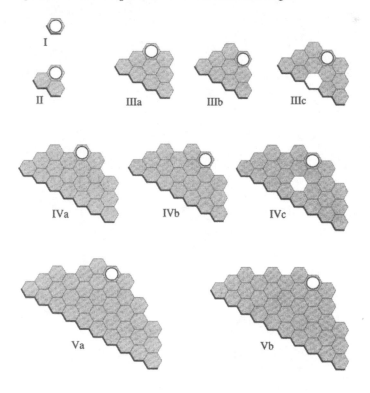

Figure 12.7 The ten minimal safe edge templates.

Figure 12.7 shows the ten safe minimal safe edge templates between a piece and its nearest home edge. It is not possible for Black to stop these pieces connecting unless they intrude into the template with a forcing move that poses an additional threat that cannot be ignored. [See Sections 5.1.1 and 5.1.2]

12.6 Forcing Moves

Forcing moves are moves to which the opponent is forced to reply on the next turn, such as moves that threaten a very strong connection or an outright win. A winning series of forcing moves is shown in Figures 12.8 and 12.9.

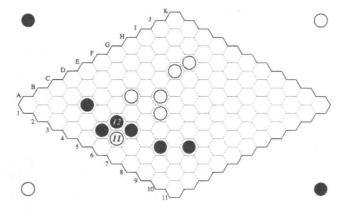

Figure 12.8. Forcing move *11*.

White's move *11* B6 is a forcing move that requires Black's reply *12* C5 to avoid immediate defeat.

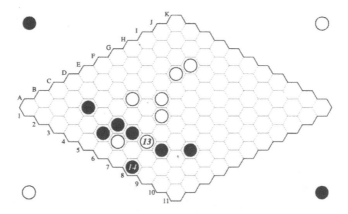

Figure 12.9 Forcing move *13* sets up the win.

White then plays forcing move *13* C7 to which Black is forced to reply *14* A8 to again avoid immediate defeat. White now has an easy win with move *15* B7.

This exchange demonstrates the importance of *momentum* within a game; in an otherwise close game, the player with the move in hand has the advantage.

The following options are available to the player when the opponent has just made a forcing move (see [Boll 1994]):

* *answer the forcing move and save the link,*

* *give up the link and move elsewhere (if not a winning link), or*

* *play a forcing move of their own.*

The appropriate choice for a given situation depends largely on the severity of the forcing move. If it is just a nuisance move that threatens to break a minor link, the forcing move may be ignored. If the forcing move is a potential game-winner, it must be answered immediately. In general *do not ignore a forcing move unless the reply is a stronger forcing move itself or the threatened connection is not essential.* [See Sections 6.3 and 7.4]

12.7 Ladders

A *ladder* occurs when a player strives to force a connection to an edge but is deflected by the opponent a constant distance away from it, resulting in a sequence of moves in a direction parallel to the edge. Figure 12.10 shows a situation in which a ladder is about to form.

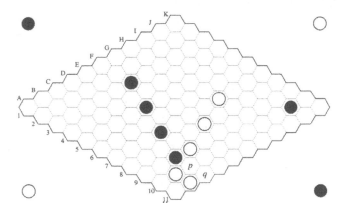

Figure 12.10. A ladder forms along the bottom right edge...

If it were White's move they could play at point *p* to win the game. However, it is Black's turn and they wisely choose the key point *p*, forcing White to block the impending connection by moving at *q*. This attack-block sequence continues along row 10, as shown in Figure 12.11, until Black reaches the stray piece at J10, which allows them to jump ahead of the ladder and win. [See Sections 7.1 and 7.2]

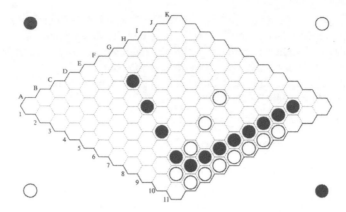

Figure 12.11. ... and Black escapes to win the game.

12.7.1 Ladder Escapes

The additional piece J10 that lies along the ladder path in the above example is described as a *ladder escape* for Black. The ladder escape is one of the most important Hex concepts and is where the game is usually won or lost. Ladders and ladder escapes are the most common way to connect piece groups in the middle of the board to each edge.

To be successful, a ladder escape should:

- *be safely connected to the target edge (usually by an edge template), and*

- *not interfere with the ladder's projected path.*

The danger of ladder escapes is one reason that the defender should strive to keep the ladder as far from the target edge as possible: the further out the ladder forms, the smaller the chance of a successful escape.

In order to determine a ladder's connection to its target edge, it's useful to recognize common ladder escape patterns. Figure 12.12 shows the simplest ladder escape formations for ladders two rows removed from the edge. Note that this set of ladder escape templates is exactly the set of edge templates. Escape templates exist for ladders further from the edge, but become fewer in number as the distance to the edge increases. [See Section 7.3, 7.6, and 7.7]

12.7.2 Forking Ladder Escapes

The surest way to ensure a ladder's escape is to make the escape piece a forcing move that threatens a more immediate threat. Many games of Hex are won with such a move. The forking ladder escape is illustrated in Figures 12.13 and 12.14.

White, whose turn it is to play, notices two points p and q at which they can breach Black's defense. As a move at q will result in a ladder along column B in the direction shown, it is in White's best interest to engineer an escape for this ladder.

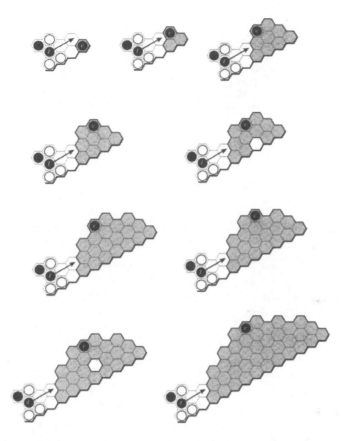

Figure 12.12. Ladder escape templates for 2-row ladders. Ladder origins are marked *l* and ladder escape pieces are marked *e*.

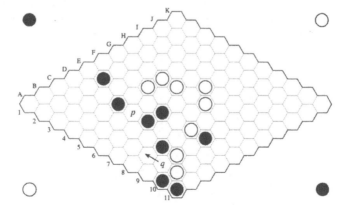

Figure 12.13. White to play.

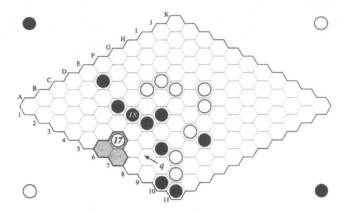

Figure 12.14. White plays a ladder escape fork with move *17* B6.

White's move *17* B6 provides the desired escape and at the same time threatens to connect through point *p* to their main body of pieces. These joint threats do not overlap at any point and form a *forking ladder escape*. Black is forced to play at *p*, conceding the ladder and the game. [See Section 7.4]

12.7.3 Ladder Escape Foils

In order to defend successfully against a ladder escape, the defender must either:

- *intrude into the ladder escape template, or*

- *block the projected ladder path.*

In the case of defending against a ladder escape fork the situation is more dire, and the defender must achieve both of these aims with the next move. This is demonstrated in Figure 12.15, where it is Black's turn to play and attempt to foil White's forking ladder

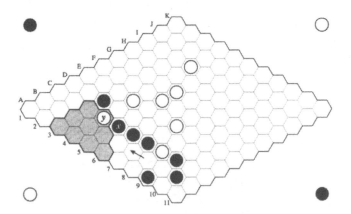

Figure 12.15. Black to play and defeat White's ladder escape.

escape at piece *y*. The direction of White's potentially winning ladder along column B is indicated by the arrow.

Notice that Black has a piece *x* adjacent to the escape piece *y* that lies on the same side of the escape piece as the ladder's point of origin. This is sufficient to *foil* the ladder as shown in Figure 12.16.

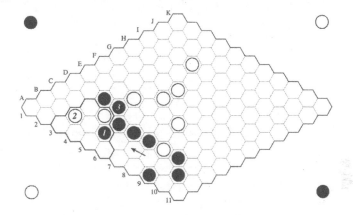

Figure 12.16. Black foils the escape.

Black's move *1* B5 foils the ladder escape and sets up a win. If White attempts to keep the threat alive by playing move *2* B4 then Black needs only reply *3* D4 to put the win beyond doubt. The key to the success of Black's ladder escape foil is that it intrudes into the potential escape's template with an adjacent move. This leaves no vulnerable points that White may later exploit as forcing moves to rekindle the threat. [See Sections 7.5 and 7.7]

12.8 Spanning Paths

The strength of any board position can be analyzed using path analysis based on:

- *connections between pieces,*

- *edge templates, and*

- *potential ladder escapes.*

Figure 12.17 shows that White has a winning path using only bridges and edge templates. If Black intrudes into this winning path at any point, White has an alternative reply that will negate the intrusion.

The shortest path between each player's edges is described as their *minimal spanning path* and indicates the strength of their position. A 0-connected spanning path provides a win for the player. It is not possible for both players to have a 0-connected spanning path.

The method for determining each player's spanning path accurately is nontrivial. Concepts used to describe the path's connectivity are introduced in Chapter 4 (Groups,

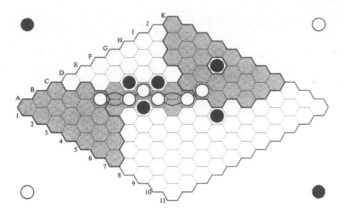

Figure 12.17. White has a winning spanning path.

Steps, and Paths) and an algorithm for performing the path generation is presented in Chapter 8 (Algorithmic Board Evaluation).

The golden rule of Hex is that *a player's position is only as good as the weakest link in their best connection across the board*, where their best connection across the board is given by their minimal spanning path. With each move the player should attempt to improve their weakest link or exploit the opponent's weakest link. Any move that achieves both of these objectives at once is a strong move.

For instance, point *p* shown in Figure 12.18 is the weakest link in the best spanning path of both players. Whoever's turn it is to play would be wise to take this key point; in fact this move is a win for either player.

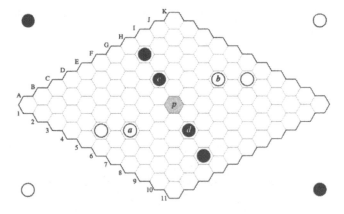

Figure 12.18. Point *p* is the weakest link in both players' strongest connection.

12.9   Multiple Threats Per Move

Just as alternative paths between pieces improve their connectivity, so too do multiple threats increase the usefulness of a move. This is most obvious in forking moves which threaten to complete two or more connections directly, but also applies to less tangible threats that may accumulate to make a move exceptional.

As the opponent only has one reply before the player's next turn, they must decide which of the multiple threats is most pressing and needs to be answered. For this reason multiple threats within a move are more effective if they are disjoint and cannot be defeated by a single reply.

Players should *strive to make every move achieve at least two parallel goals or threats, direct or implied.*

The player can optimize this strategy by choosing moves that hide implied threats behind more obvious ones. There is the chance that the opponent will blindly reply to the greater threat and fail to notice or block the more subtle one, which may return to hurt them later in the game. [See Section 10.1]

12.10   Looking Ahead

It is often difficult to determine which player has the upper hand in a given position, unless the game is a win for either player (0-connected spanning path). The best way to accurately determine the state of play is to perform a lookahead search to resolve any undecided regions on the board. Such searches fall into two broad categories:

• *Short term lookahead*: An exhaustive analysis of play within a limited region: for example, whether a given piece can safely connect to its nearest home edge. This approach is 100% accurate (providing that no mistakes are made!) but becomes impractical for all but short sequences of moves.

• *Long term lookahead*: A projection of possible play requiring some assumptions be made as to which move the opponent will choose at key points. This approach is not 100% accurate and becomes less precise with each assumption made. It is useful for predicting patterns of play that may emerge in several moves' time in order for the player to prepare their attack/defense accordingly.

There is no easy solution to Hex. Between two otherwise equally skilled opponents, the player who is willing to work harder and perform the more thorough lookahead will usually win the game. [See Section 10.4]

Hex Puzzles

This chapter contains a collection of previously published Hex puzzles as well as several original problems designed to supplement the points of strategy discussed in previous chapters. The puzzles presented are a combination of abstract (devised with a specific point in mind) and practical (derived from actual situations encountered during play).

For the purpose of the following puzzles, a player is deemed to have won the game as soon as they have established a 0-connected spanning path. It is not necessary to complete a chain connecting their sides of the board as this is just a matter of filling in the gaps along the winning path, and is redundant. For some puzzles more than one move is required for a complete solution.

13.1 Previously Published

The following puzzles were collected from previously published sources.

Puzzle 1: White to move and win. From Piet Hein's original series of articles [1942].

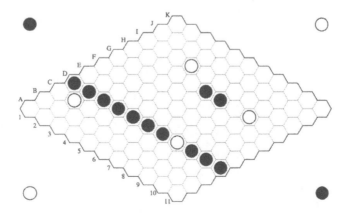

Puzzle 2: White to move and win. Devised by Piet Hein, published in [Gardner 1959].

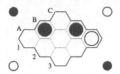

Puzzle 3: White to move and win. Devised by Piet Hein, published in [Gardner 1959].

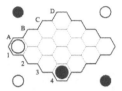

Puzzle 4: White to move and win. Devised by Piet Hein, published in [Gardner 1959].

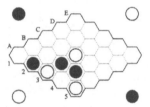

Puzzle 5: Black to move and win? Originally proposed by Claude Berge [1981].

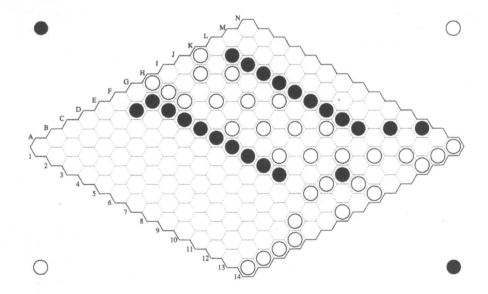

Puzzle 6: Black to move and win (easy). Designed by Bert Enderton [1995].

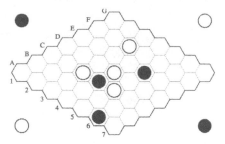

Puzzle 7: Black to play and win (medium). Devised by Bert Enderton [1995].

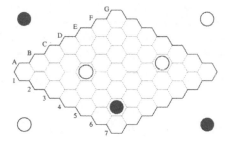

Puzzle 8: White to move and win (hard). Devised by Bert Enderton [1995]. Note: this puzzle is deceptively difficult.

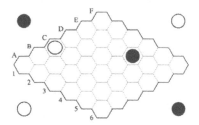

Puzzle 9: White to move and win (Reverse Hex rules). Reverse Hex is a variant of Hex where the first player to force their opponent to connect their edges wins the game. Composed by Ron Evans and reprinted in [Gardner 1988a].

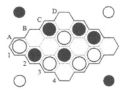

Puzzle 10: Where should White move: *a, b,* or *c*? This puzzle was originally published with respect to the game of Y [Schensted and Titus 1975]. However, the principles involved apply equally to Hex.

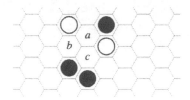

13.2 Original

The following puzzles were created specifically to illustrate points of strategy discussed in this book; hence, they reward analysis based on gameplay principles rather than exhaustive search. Many were derived from actual games to give the reader a feel for particular strategies within the context of the overall game, but handcrafted to ensure that they are:

- *Correct:*   *the solution cannot be refuted by an alternative line of play.*

- *Concise:*   *the solution given is the simplest.*

- *Unique:*   *there are no other equally good solutions.*

Puzzle 11: Are these two pieces safely connected to the edge?

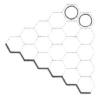

Puzzle 12: Is this piece safely connected to the edge?

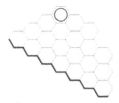

Puzzle 13: What is Black's best move?

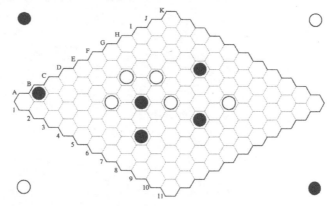

Puzzle 14: What is Black's best move?

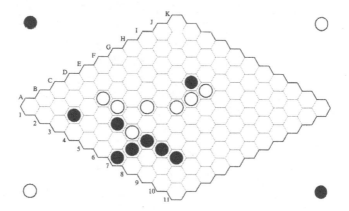

Puzzle 15: Black to play and win.

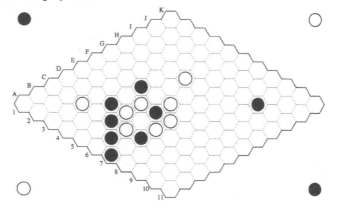

Puzzle 16: White to move and avoid immediate defeat.

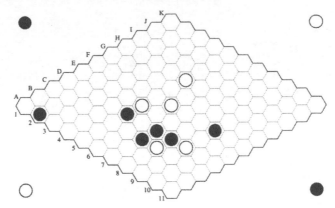

Puzzle 17: White to play and win.

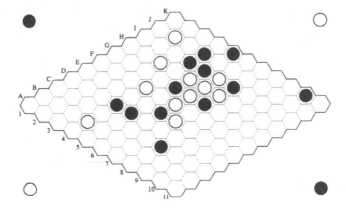

Puzzle 18: White to play and win.

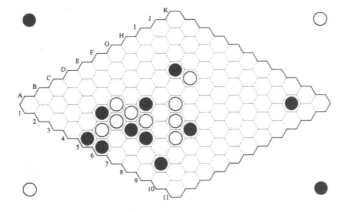

Puzzle 19: Black to play and win.

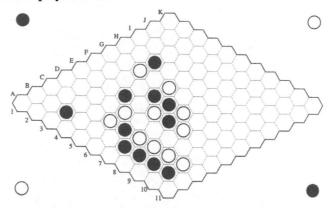

Puzzle 20: White to play and win.

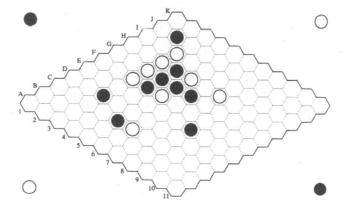

Puzzle 21: Black to move and block White's impending connection.

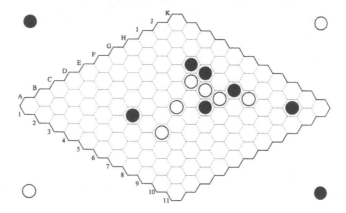

Puzzle 22: Black to play. What is White's most vulnerable point?

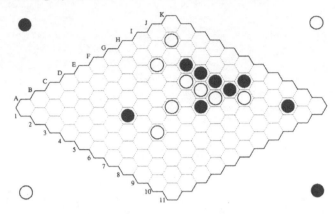

Puzzle 23: Black to play and win.

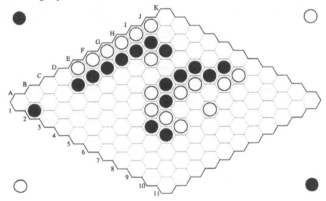

Puzzle 24: White to play. What move is their only sensible choice?

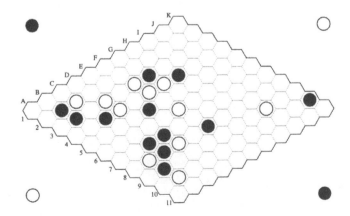

Puzzle 25: White to play and win.

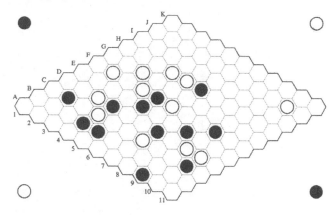

Puzzle 26: White to play and win.

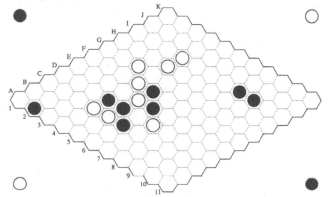

Puzzle 27: Black to play and win.

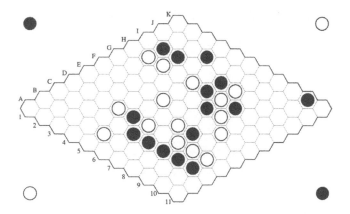

Puzzle 28: White to play and win.

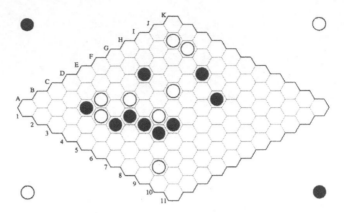

Puzzle 29: White to play and win.

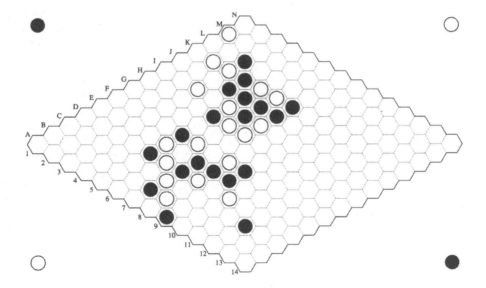

Puzzle 30: Black to play and win. Puzzle designed by Leonid Gluhovsky.

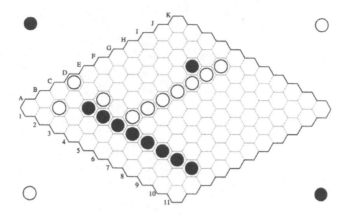

Conclusion

The reader who has understood all of the material covered in *Hex Strategy* should have the knowledge required to play Hex at a high level. However, this knowledge does not eliminate the complexity of the game—it reduces the effort required to analyze a board position and indicates worthwhile traps or killer moves to be made, but does not provide a solution to Hex. The player prepared to work harder and explore a greater number of possible continuations of play will prevail over a lazy player of equal skill.

The reader who overlooked some aspects of strategy, or simply wants to learn the game gradually, should refer to *Hex Strategy* from time to time. They may recognize patterns of play from their own games that crystallize ideas discussed in the book, and make the reading of more advanced topics that much easier.

It's recommended that players practice as often as possible in order to see the strategies discussed actually put into action on the Hex board. If a playing partner is not readily available, a good Hex program such as Hexy by Vadìm Anshelevich [1999] will provide ample opportunity to learn the game.

One of the aims of this work was to provide concise definitions for points of strategy that players use regularly and instinctively but cannot quite put into words. This will hopefully provide a common ground for further discussion of Hex strategy and promote future interest and research into the game.

References

The following is a list of publications cited in this book, and other material relevant to the study of Hex. Web page references and other online resources are listed in the second section.

15.1　Publications

Abbott, R. (1988) "What's Wrong With Ultima", *World Game Review*, 8, 12-13.

Adams, C. (1994) *The Knot Book*, W. H. Freeman, New York.

Aizawa, K. and Nakamura, A. (1989) "Grammars on the Hexagonal Array", *International Journal of Pattern Recognition and Artificial Intelligence*, 3:3/4, 469-477.

Alpern, S. and Beck, A. (1991) "Hex Games and Twist Maps on the Annulus", *American Mathematical Monthly*, 98, 803-811.

Arnold, P. (1975) *The Illustrated Book of Table Games*, St. Martin's Press, New York.

Arnold, P. (1985) *The Book of Games*, Golden Press, Sydney.

Bailey, D. (1997) *Trax Strategy for Beginners*, D. G. Bailey, Palmerston North. Second edition.

Beasley, J. D. (1989) *The Mathematics of Games*, Oxford University press, Oxford, 141-143.

Beck, A. (1969) "Games", *Excursions Into Mathematics*, Eds. Beck, A., Bleicher, M., and Crowe, D., Worth, New York, 317-387.

Berge, C. (1981) "Some Remarks about a Hex Problem", *The Mathematical Gardner*, Ed. Klarner, D. A., Wadsworth, Belmont, 25-27.

Berlekamp, E. R., Conway, J. H., and Guy, R. K. (1982) *Winning Ways For Your Mathematical Plays*, Academic Press, London, 218-219, 679-682.

Berman, D. (1976) "Hex Must Have a Winner: An Inductive Proof", *Mathematics Magazine*, 49:2, 85-86. This issue also includes two short responses to Berman's article by Paul Johnson and Daniel Zwillinger, 156.

Berloquin, P. (1976) *100 Perceptual Puzzles*, Barnes and Noble, New York.

Binmore, K. (1992) *Fun and Games: At Text on Game Theory*, D. C. Heath, Lexington.

Book, D. L. (1998) "What the Hex", *The Washington Post*, September 9th, page H02.

Broline, D. (1981) "Kriegspiel Hex", *Mathematics Magazine*, 54:2, 85-86 (solution).

Coxeter, H. (1973) *Regular Polytopes*, Dover, New York. Third edition.

Davis, M. (1976) "On Artificial Machine Learning: Some Ideas in Search of a Theory", *International Journal of Computer Mathematics*, B:5, 315-329.

Davis, M. (1986) "Computer Learning of Parlor Games", *Physica*, 22D, 351-354.

Dresher, M. (1961) *Games of Strategy: Theory and Applications*, Prentice Hall, Englewood Cliffs.

Evans, R. (1974) "A Winning Opening in Reverse Hex", *Journal of Recreational Mathematics*, 7:3, 189-92. Reprinted in *Mathematical Solitaires & Games*, Ed. Schwartz, B., Baywood, Amityville, 104-107.

Evans, R. (1975-76) "Some Variants of Hex", *Journal of Recreational Mathematics*, 8:2, 120-22.

Even, S. and Tarjan, R. (1976) "A Combinatorial Problem Which Is Complete In Polynomial Space", *Journal of the ACM*, 23, 710-19.

Fraenkel, A. (1996) "Combinatorial Games: Selected Bibliography with a Succinct Gourmet Introduction", in *Games of No Chance*, ed. Nowakowsi, R., Cambridge University Press, Cambridge, 493-537.

Gale, D. (1979) "The Game of Hex and the Brouwer Fixed-Point Theorem", *The American Mathematical Monthly*, 86:10, 818-827.

Gardner, M. (1959) "The Game of Hex", *Mathematical Puzzles and Diversions*, Penguin, Hammondsworth, 70-77.

Gardner, M. (1961) "Recreational Topology", *More Mathematical Puzzles and Diversions*, Penguin, London, 60-68.

Gardner, M. (1966) "Bridg-It and other Games", *New Mathematical Diversions from Scientific American*, Simon and Schuster, New York, 210-218.

Gardner, M. (1988a) *Time Travel and Other Mathematical Bewilderments*, W. H. Freeman, New York, 158-159.

Gardner, M. (1988b) "Afterword, 1988", *Hexaflexagons and Other Mathematical Diversions,* University of Chicago Press, Chicago, 182-184. This book is a revised print of Gardner's *Mathematical Puzzles and Diversions.*

Gasser, R. (1996) "Solving Nine Men's Morris", in *Games of No Chance*, Ed. Nowakowsi, R., Cambridge University Press, Cambridge, 101-113.

Ghyka, M. (1977) *The Geometry of Art and Life*, Dover, New York.

Guy, R. K. (1991) *Combinatorial Games*, American Mathematical Society, Providence, 143.

Headington, R. (1973) "The Game of Hex", *Games & Puzzles*, 16, 8-9.

Hein, P. (1942) "Polygon", *Politiken*, December 26th.

How, M. (1984) "Beyond Hex", *Games*, July, 57-58.

Jacoby, O. and Crawford, J. (1970) *The Backgammon Book*, Macmillan, London.

Johnson, P. and Zwillinger, D. (1976) "Simpler, Simpler...", *Mathematics Magazine*, 49:2, 156. Two letters printed in the "News and Letters" section in response to David Berman's article of the same issue.

Keller, M. (1998) "A Few Definitions?", *World Game Review*, 13, 3.

Koch, K. (1991) "A Winding Road", *Pencil & Paper Games*, 49-51.

Lehman, A. (1964) "A solution to the Shannon switching game", *SIAM Journal of Applied Mathematics*, 12, 687-725.

Levy, D. (1983) *Computer Gamesmanship*, Simon and Schuster, New York.

Mallett, J. (1998) "The Game of Y", *World Game Review*, 13, 17-18.

McCallion, J. (1992) "Pipeline", *World Game Review*, 11, 26.

McConville, R. (1974) *The History of Board Games*, Creative Publications, Palo Alto.

McWorter, W. (1979) "Kriegspiel Hex", *Mathematics Magazine*, November, Problem 1084 (initial proposal).

Midgley, R. (1975) *Waddington's Illustrated Encyclopedia of Games*, Pan, Sydney.

Nasar, S. (1994) "The Lost Years of a Nobel Laureate", *New York Times*, November 13th.

Nasar, S. (1998) *A Beautiful Mind*, Simon and Schuster, New York.

Nowakowsi, R. (1996) *Games of No Chance*, Cambridge University Press, Cambridge.

O'Rourke, J. (1998) *Computational Geometry*, Second Edition, Cambridge University Press, Cambridge.

Pappas, T. (1994) *The Magic of Mathematics*, Wide World, San Carlos.

Parlett, D. (1999) *The Oxford History of Board Games*, Oxford University Press, Oxford.

Pickover, C. (1990) "Picturing Randomness with Truchet Tiles", *Computers, Pattern, Chaos and Beauty: Graphics From an Unseen World*, St. Martin's Press, New York, 329-332.

Pierce, J. R. (1961) *Symbols, Signals and Noise*, Harper's, New York. Reprinted in 1980 as *An Introduction to Information Theory: Symbols, Signals and Noise - Second, Revised Edition*, Dover, New York.

Schensted, C. and Titus, C. (1975) *Mudcrack Y & Poly-Y*, NEO Press, Peaks Island.

Schmittberger, W. (1983) "Star: A Game Is Born", *Games*, September, 51-54.

Shannon, C. (1953) "Computers and Automata", *Proceedings of the Institute of Radio Engineers*, Vol. 41.

Stoutamire, D. (1991) *Machine Learning, Game Play, and Go*, Center for Automation and Intelligent Systems Research TR 91-128, Case Western Reserve University.

Straffin, P. (1985) "Three Person Winner-Take-All Games with McCarthy's Revenge Rule", *The College Math. Journal*, 16:5, Nov, 386-394.

Tesauro, G. (1995) "Temporal Difference Learning and TD-Gammon", *Communications of the ACM*, 38: 3, 58-68.

Thomashow, M. (1986) "Octiles", *World Game Review*, 11,27.

Thrun, S. (1995) "Learning to Play the Game of Chess", *Advances in Neural Information Processing Systems 7*, Eds. Tesauro, G., Touretzky, D., and Leen, T., MIT Press.

Tosic, R., Masulovic, D., Stojmenovic, I., Brunvoll, J., Cyvin, B., and Cyvin, S. (1995) "Enumeration of Polyhex Hydrocarbons to $h = 17$", *Journal of Chemical Information and Computer Sciences*, 35, 181-187.

Walker, S., Lister, R. and Downs, T. (1993) "On Self-Learning Patterns in the 'Othello' Board Game", *Proceedings of the 6th Australian Joint Conference on Artificial Intelligence*, World Scientific, Singapore, 328-333.

Wang, Y. and Bhattacharya, P. (1997) "Digital Connectivity and Extended Well-Composed Sets for Gray Images", *Computer Vision and Image Understanding*, 68:3, 330-345.

Wells, D. (1973) "Twixt", *Games & Puzzles*, 1:18, October, 4-7.

Weisstein, E. (1999) *CRC Concise Encyclopedia of Mathematics*, CRC Press, Boca Raton, p835.

Weyl, H. (1952) *Symmetry*, Princeton University Press, New Jersey.

15.2   Online Resources

The following Internet references were current at the time of writing (April 2000).

Anshelevich, V. (1999) "Hexy Plays Hex", http://home.earthlink.net/~vanshel/.

Anshelevich, V. (2000) "The Game of Hex: An Automatic Theorem Proving Approach to Game Programming", http://home.earthlink.net/~vanshel/.

Bailey, D. (1999) "Trax: Frequently Asked Questions", http://www.traxgame.com/faq/faqb.html.

Bergstrom, F. (1998) "Fawna's Hex Lessons", http://angel.pomona.edu/Academics/Course/chess/wwwlinks.html.

Boll, D. (1994) "Hex: Answers to common questions", http://www.frii.com/~dboll/hexfaq.txt.

Boll, D. (1997) "Hex, Hex FAQ, comments", posting to newsgroup *rec.games.abstract*, 3rd April.

Bourke, P. (1996) "Hexagonal Lattice", http://www.mhri.edu.au/~pdb/geometry/hexagon/.

Brasa, E. (1999), "Hex game information center", http://www.microring.it/hex/.

Enderton, H. (1995) "Answers to infrequently asked questions about the game of Hex", http://www.cs.cmu.edu/~hde/hex/hexfaq.

Finn, R. (1998) "Hex by Piet Hein", http://members.iex.net/~rfinn/gameshlf/abstract/hex/ hex.htm.

Freeling, C. (1999) "The Arena", http://www.mindsports.net/Arena/.

Frogley, T., Fenwick, N., Lester, M., and Norton, D. (1998) "Thoughtwave: A quick-moving tactical contest for 2 quick-thinking players", http://www.geocities.com/TimesSquare/ Stadium/2652/.

Go2Net (1999) "PlaySite", http://www.playsite.com/.

King, D. (1999) "The Game of Hex", http://homepages.enterprise.net/drking/hexagons/hex/.

Krammer, T. (1999a) "The abstract board game Hex", http://huizen.dds.nl/~krammer/hex.htm.

Krammer, T. (1999b) "The abstract board game Trinidad", http://huizen.dds.nl/~krammer/ trinidad.htm.

Lycos Games (1999) "Hex", http://playsite.lycos.com/games/board/hex/.

Maitreg (1999) "Hex!", http://www.maitreg.com/hex/index.html.

MazeWorks (1999) "Hex-7", http://www.mazeworks.com/hex7/hex7.htm.

O'Sullivan, S. (1998) "Kaliko", http://www.io.com/~sos/bc/kaliko.html.

Peterson, I. (1997) "Silicon Champions of the Game", *Science News Online*, Feature Article, 8th February, http://www.sciencenews.org/sn_arc97/8_2_97/bob1.htm.

Rognlie, R. (1996) "Welcome to Richard's Play-By-eMail Server (Gamerz.NET)", http://www.gamerz.net/~pbmserv/.

Rusin, D. (1998) "The Game of Hex", http://www.math.niu.edu/~rusin/papers/uses-math/games/hex/.

Schensted, C. and Titus, C. (1998), "The Game of Y", http://www.gamepuzzles.com/goyh.htm.

Steere, M. (1996) "Help for Hexbo", http://www.gamerz.net/~pbmserv/hexbo.html.

Van Rijswijck, J. (1998) "Hex", http://www.cs.ualberta.ca/~javhar/hex/.

Van Rijswijck, J. (1999) "Queenbee's Home Page", http://www.cs.ualberta.ca/~queenbee/.

Van Rijswijck, J. (2000) "Are Bees Better Than Fruitflies? Experiments with a Hex Playing Program", http://www.cs.ualberta.ca/~javhar/.

Solutions to Puzzles

Solution 1: F7. White has a safe non-overlapping path to the right side and a ladder escape guaranteeing connection to the left.

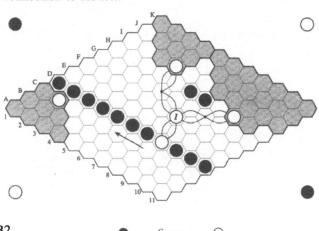

Solution 2: B2.

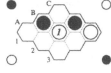

Solution 3: B3.

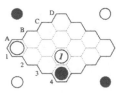

Solution 4: C2.

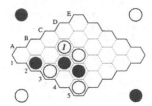

Solution 5: See Appendix B.

Solution 6: B4. White's defense *2* D1 can be beaten by the sequence shown below.

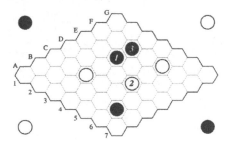

Solution 7: E3. White's best defense is *2* D5 which is defeated by *3* F3.

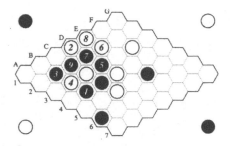

Solution 8: D4. This puzzle demonstrates the extraordinary complexity of Hex analysis even for the small 6x6 board.

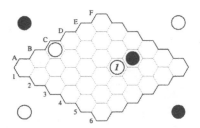

Proof of this solution is nontrivial, and many experienced players will give up before solving it. Some of Black's most promising lines of play and their refutation by White are shown below.

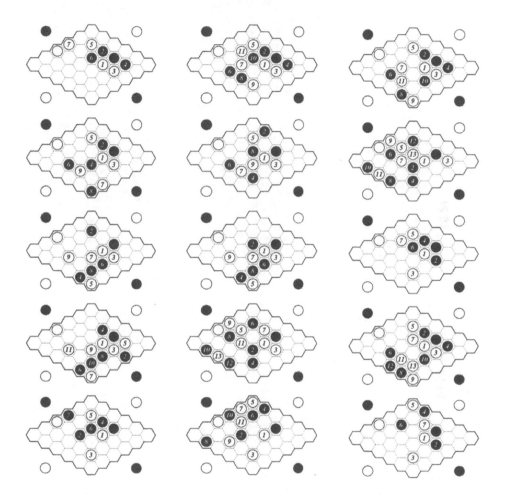

Now consider this level of complexity increased exponentially for the 11x11 and larger boards to get a feel for the difficulty of solving the game at these sizes!

Solution 9: B1. If White plays any other move, then Black can force the win by taking B1.

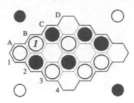

Solution 10: Move *b* is best. All three moves connect White's pieces, but *b* gains extra territory at point *p*. This advantage is subtle but could mean the difference between a win and a loss. Black can't exploit the bridge's vulnerable points with a forcing move to gain territory themselves; the link is as good as an adjacent connection.

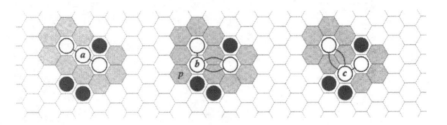

Solution 11: This is the only move that defeats the connection. It forms a III template to each side of the region which White is unable to cross.

Solution 12: Black can block the connection with the following sequence of moves. Move *1* forces White to attempt a 1-row ladder escape with move *4*, but Black is able to cut this escape off with template intrusion *5* followed by the key move *7*.

Solution 13: White almost has a 0-connected spanning path, but not quite. Point *p* is vulnerable as two edge templates overlap there. Leonid Gluhovsky points out that White can win on the next move by playing at J2.

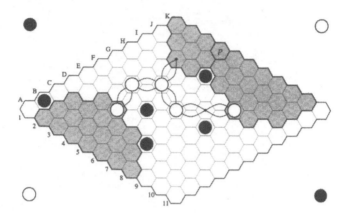

Solution 14: D5 is the only vulnerable point. The other shaded region near B4 may appear vulnerable, but any move there can be defeated easily by White.

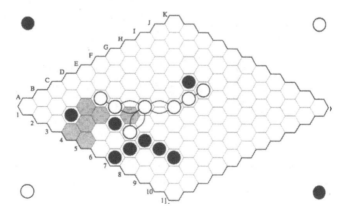

Solution 15: B8. White cannot stop Black from forming a ladder which will eventually connect with escape piece I9. Black is safely connected to the top left edge (the reader may wish to prove this for themselves).

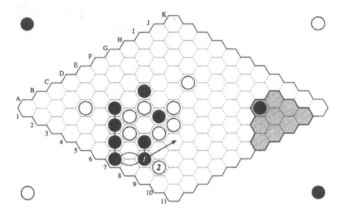

It may appear that *1* C8 is a more direct block for Black. However, this gives White the opportunity to play the killer move *2* B10. This is a forking ladder escape which gives White the win.

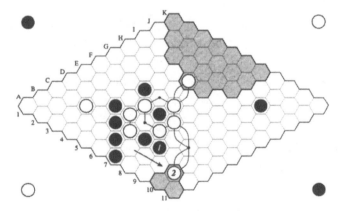

It's clear that Black must move somewhere in column B to block the ladder. How about *1* B9, which both blocks the ladder and extends Black's connection with a bridge? Unfortunately, this is open to forcing move *2* B8 which again gives White the win.

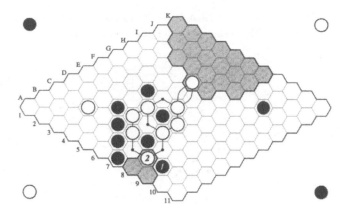

Solution 16: E10. If White plays the obvious block *1* E8, then Black can force a ladder and win.

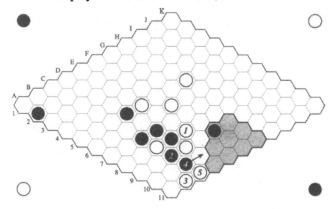

Instead White must play a move that both blocks the ladder path and intrudes into Black's template. E10 is such a move. Black is able to reconnect their broken template with move *2*, but White shuts the ladder out with *3*.

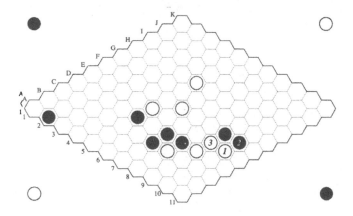

A better reply by Black is *2* E9, which threatens to connect across both sides of piece *1*. However, White is again able to shut the ladder out with *3* D10.

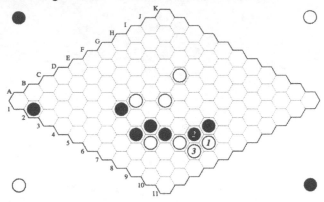

Solution 17: D6. This forms a single group of White pieces connected to the top right edge...

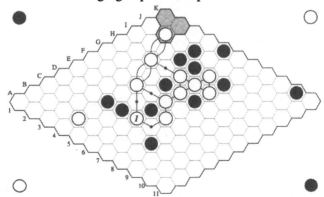

... which Black is unable to stop connecting to the bottom left edge as well, due to the ladder escape at B4.

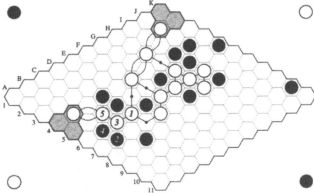

This is a good example of the need to consider connections based on groups, not just chains, when performing path evaluation.

Solution 18: B8. This is a standard forking ladder escape.

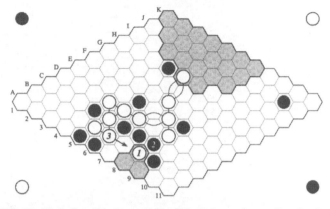

Solution 19: D6. The fact that this is a win for Black is not immediately obvious until one realizes that piece *y* is safely connected to the top left edge without interfering with piece *x*'s edge template.

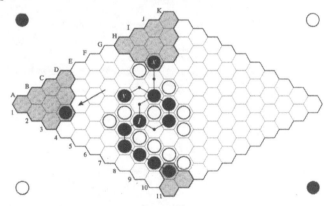

Solution 20: D4. It may appear that White's connection is vulnerable due to the overlap at point *p*...

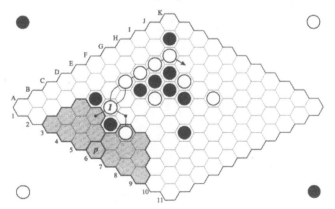

... however, this cannot be successfully exploited by Black.

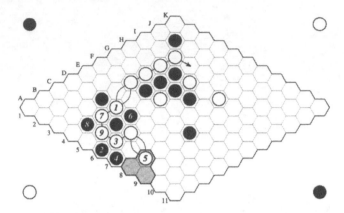

The 3-row ladder down column I can safely connect to the top right edge but requires the use of both escape pieces H5 and H7.

Solution 21: J7. The more obvious block *1* H8 can defeated by White by forcing a ladder with move *2* I6.

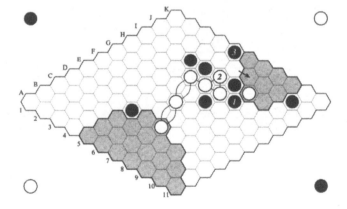

1 J7 is a classic case of a ladder escape foil played adjacent to the escape piece.

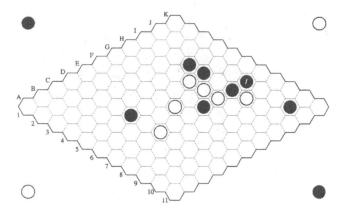

Solution 22: G5 is the most vulnerable point in White's spanning path. White is in a very strong position and should win shortly.

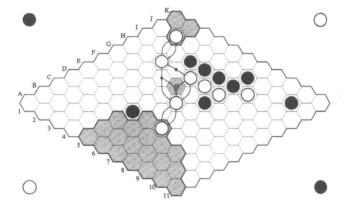

Solution 23: C6. Move *1* D5 looks promising for Black but allows White to play killer move *2* B10, which gives them a forking ladder escape and just about sets up a White victory.

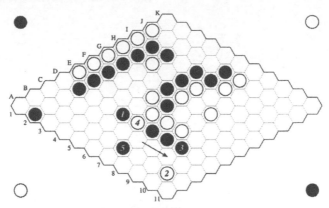

Move *1* C6 is far superior and guarantees the win for Black.

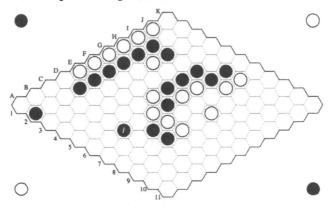

Solution 24: C3. Black has two 1-connected spanning paths. The first spanning path uses a cascading ladder escape to the bottom right edge and joins to the left corner.

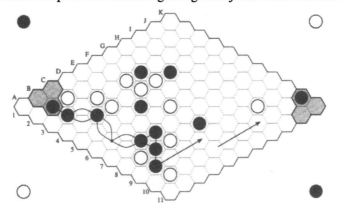

The second spanning path requires a ladder along row 2 that also joins to left corner.

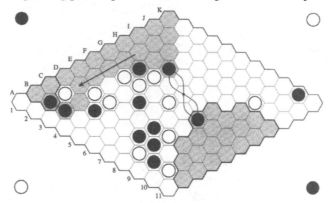

C3 is the only vulnerable point at which these paths overlap.

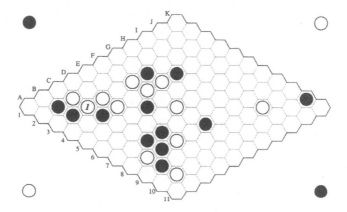

Solution 25: C5. This is obviously the most vulnerable point in Black's best spanning path.

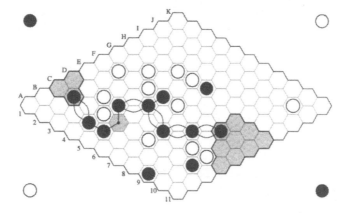

Black has two basic lines of defense in this situation: *2* B6 and *2* A7. However, defense *2* B6 fails...

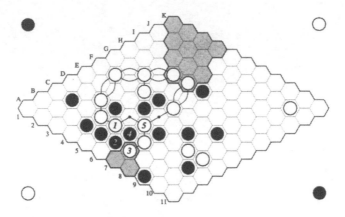

... as does defense *2* A7.

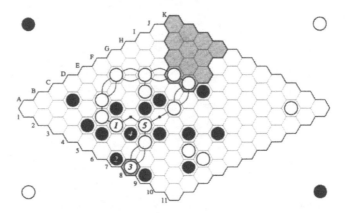

Note the use of the two-piece edge template discussed in Section 5.3.

Solution 26: D8. This forms a two-piece template with the bottom left edge, and Black is unable to stop White's connection across the board.

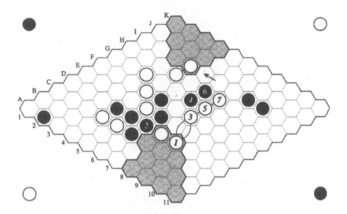

The more obvious connection *1* D6 ends badly for White, as Black is able to force a ladder along row 8, 9, or 10 for the win.

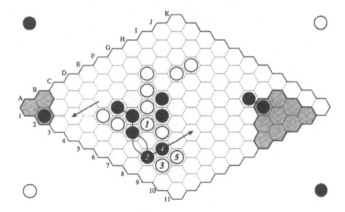

If White tries instead to block with move *1* C7 then Black is able to force a ladder from above along row 7 for the win.

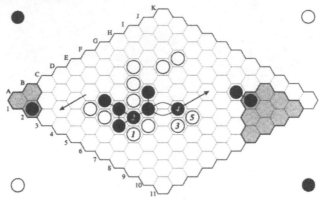

Solution 27: F6. White is forced to reply *2* E6. After Black plays *3* F4, then White must block their passage to the top, for which *4* F3 is probably the best move. However *5* G4 gives Black the win next move.

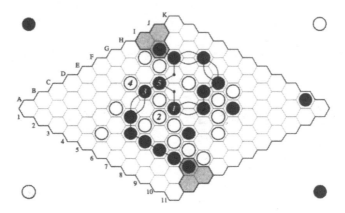

Solution 28: G4. This gives White an unbeatable spanning path across the board.

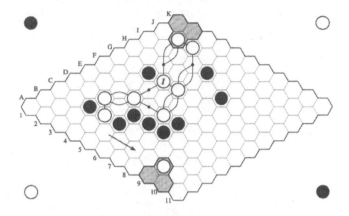

If White plays the more obvious defense *1* F5, then Black can complete a spanning path using template Va to the bottom right edge.

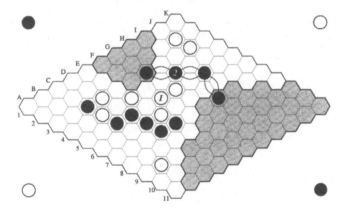

If White instead blocks this path with a move such as *1* H4, then Black is able to connect to their pieces D7 and E7 which White cannot block from reaching the bottom. White's original *1* G4 intrudes on both of Black's potential winning paths.

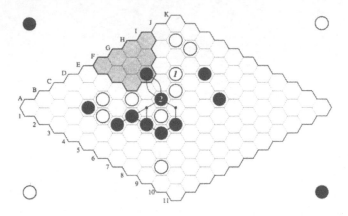

Solution 29: G4. First consider White's chain *a* shown below. Black is unable to block its connection to the edge (White's key move is *2* B10):

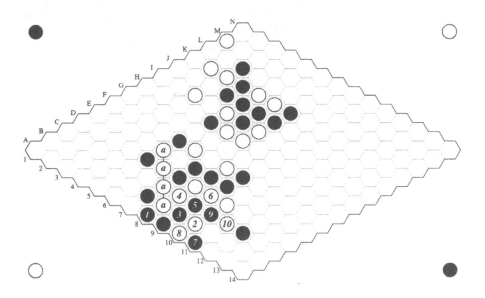

Now consider White's piece *b* as shown below. Again, Black is able to block its connection to the edge (White's key move here is *6* M5):

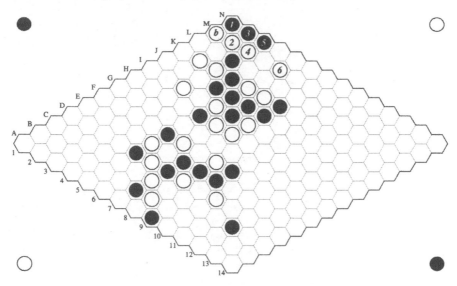

White's move G4 safely connects chain *a* with piece *b* following the sequence of moves shown below. These are forcing moves to which Black does not have any better reply.

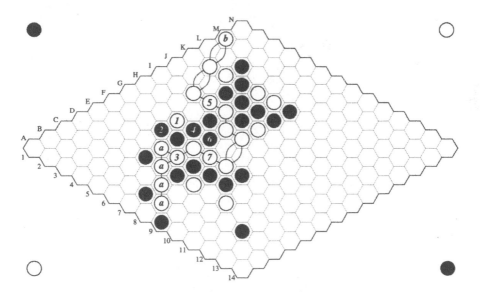

This puzzle is similar to puzzle *28*, but was drawn directly from an actual game without modification. Thanks to Kevin Walker for pointing out the winning play.

Puzzle 30: D2. Black forces a ladder along row 2 and plays the killer move *11* J2, forcing a foldback ladder along row 4 back to the main body of Black pieces.

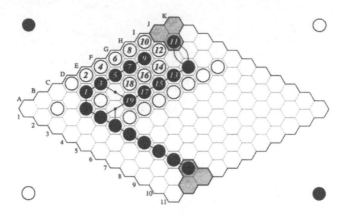

If Black attempts the alternative route through point D4, then White is able to block the connection and win the game with move *8* H2.

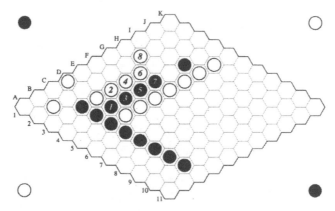

This solution is of particular interest as it:

• *demonstrates that ladder principles can apply against a solid line of defense and not just the edge, and*

• *reverses the usual trend of ladders as it travels away from the edge and towards the connecting body of pieces.*

Some Notes on Berge's Hex Problem

In 1981 Claude Berge proposed the following Hex problem: given the board position illustrated below, determine a winning play for Black [1981]. This problem is especially interesting due to the fact that White is in such a strong position and has all but won the game. Play occurs on a 14x14 board.

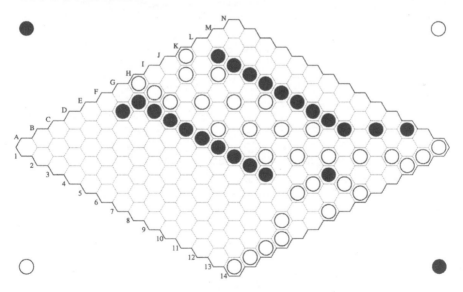

Figure B.1. Berge's Hex Problem: Black to play and win?

It was Berge's desire to develop a problem whose solution was paradoxical and unexpected. Despite the cleverness of the proposed solution, however, further analysis reveals that Black can *not* win from this position and the problem has no solution.

The fact that there are a substantially different number of pieces on the board for each player (twenty-two Black pieces and thirty-one White pieces) is not relevant to the problem. This position may have legitimately arisen during handicap play.

Figure B.2 shows path analysis of the above position using the algorithm described in Chapters 4 and 8. Safe groups <(a, b){p, q}> and <(c, d){r, s}> are shown. The two best spanning paths for White connect these two groups, and their links are shown superimposed upon each other. Grey hexagons indicate points p, q, r, s, and t where the two paths overlap. Points p, q, r, and s occur within safe groups so are not themselves vulnerable, but can be exploited as forcing moves. However, note that both spanning paths interfere at point t which is indeed vulnerable.

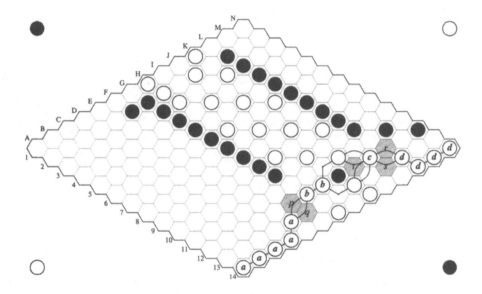

Figure B.2. Path analysis of the problem, with vulnerable points highlighted.

A number of additional spanning paths of equal connectivity exist; however, the two shown have the smallest empty point sets and are therefore the optimal spanning paths. There is no combination of 1-connected spanning paths on this board without at least one point of interference.

These points of overlap are precisely the points that Berge exploits in his proposed solution. The first few moves of the solution are illustrated in Figure B.3 and proceed as follows:

Black   White

54 F13   Forcing move that steals territory for Black and provides a ladder escape for piece *x*.

Black   White

	55 F12	Forced reply that prevents move *54* from connecting with Black's lower body of pieces along column G.
56 K13		Another forcing move that steals territory for Black and provides a potential ladder escape for existing piece *x* in the other direction.
	57 L12	Forced reply that prevents move *56* from connecting with Black's upper body of pieces along column L.
58 J12		Forcing move that links existing point *x* with both ladder escapes, at the same time threatening to connect with Black's upper body of pieces from which it is one step away.
	59 K11	Forced move that prevents *58* from connecting with Black's upper body of pieces.

This sequence results in the board position illustrated in Figure B.3.

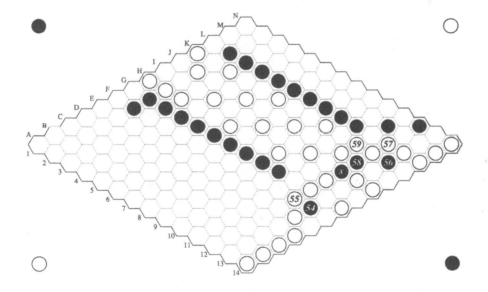

Figure B.3. Forcing moves that set up Berge's solution.

From here, Black continues to play a sequence of forcing moves that White cannot deflect, alternately threatening to connect with their lower body of pieces along column G and their upper body of pieces along column L with alternating moves. Black forces a passage through this corridor until it forms a solid connection with the top left edge. Move *84* give Black the win.

The complete winning sequence proposed by Berge is shown in Figure B.4.

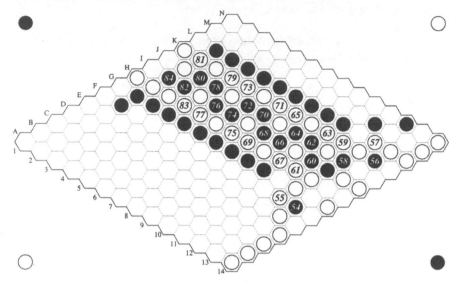

Figure B.4. Berge's complete solution.

However, consider what happens if White replies to Black's initial *54* G11 with move *55* F13 instead, as illustrated below:

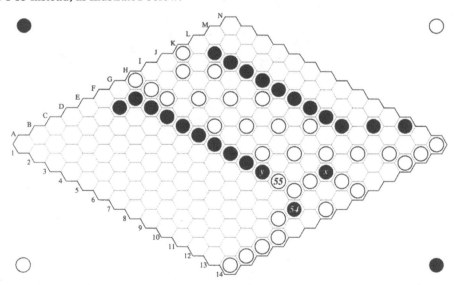

Figure B.5. Foiling move *55* interferes with the solution.

55 F13 is a foiling move that successfully answers forcing move *54* and also interferes with the connection between Black's existing pieces *x* and *y*. This becomes critical shortly.

Black's best play in this situation is to follow the same moves suggested by Berge. This time, however, Black's move *60* I11 does not threaten to connect to Black's lower main body of pieces and is therefore not a forcing move. White is free to play *61* J10 which stops Black's progress along the corridor. This position is illustrated in Figure B.6 below.

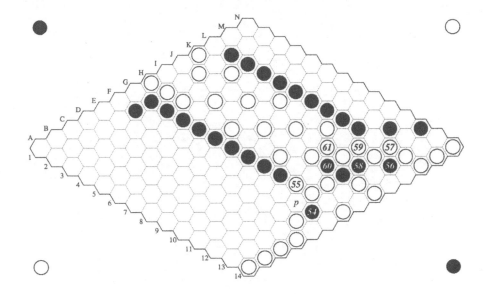

Figure B.6.   Black cannot win.

Black's last remaining hope of winning is to force a connection through point *p*. This is now the only vulnerable point along White's spanning path, so Black is in fact forced to play there. However, White is able to block Black's attempt to break through and win the game with a reasonably straightforward defense, which the reader may wish to verify for themselves.

Contrary to Berge's stipulation, Black cannot win from this position and the problem has no solution.

Please note that this discussion is not intended as a criticism of Berge's work on Hex or his proposed problem, which is indeed an interesting exercise in Hex analysis. Instead it should be seen as a reminder of the complexity of the game and the difficulty of developing nontrivial Hex problems.

Sample Games

The following games are drawn with permission from a database of games played on Richard Rognlie's Gamerz.NET server [Rognlie 1996]. The names of individual players are not shown, but those involved have been acknowledged elsewhere.

Games were selected based on the following criteria:

- *both players' ratings within the top 5% of Hex players on the server,*

- *the game is played to completion (0-spanning path or obvious win for either player), and*

- *the game shows some points of strategic interest.*

Black (denoted V for *Vertical*) starts all games in keeping with the convention used on the Gamerz.NET. Unfortunately, games stored on the server do not record whether the second player enforced the swap option, or whether the swap option is even provided for a given game.

A selection of 14x14 games is also provided to give players a feel for the game on the larger board.

In each game the last player to move is the winner.

Sample Game 1:

Black	White
1 K1	*2* F6
3 G6	*4* I5
5 H4	*6* G7
7 F7	*8* E9
9 D8	*10* C9
11 F9	*12* E10
13 I7	*14* H9

Black   **White** (Sample Game 1)
15 D9   *16* H3
17 C10   *18* I3
19 F4

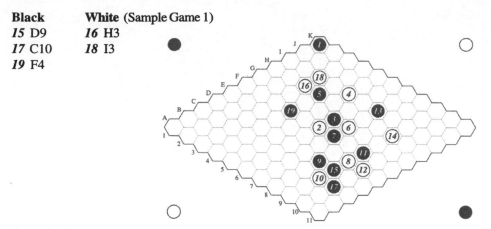

Sample Game 2:

Black	**White**
1 B3	*2* F6
3 D7	*4* C9
5 E8	*6* E9
7 D9	*8* C11
9 F10	*10* H5
11 H7	*12* D5
13 E5	*14* F3
15 E4	*16* E3
17 D4	*18* D1
19 C2	*20* D10
21 F8	

Sample Game 3:

Black	**White**
1 K10	*2* F6
3 H5	*4* H6
5 G6	*6* H4
7 J3	*8* J2
9 I3	*10* I2
11 C7	*12* C9
13 D8	*14* D5
15 H3	*16* I9 .
17 C4	*18* I4
19 G5	*20* G4

Black   **White** (Sample Game 3)
21 F5   *22* F8
23 D9   *24* E3
25 F2

Sample Game 4:

Black	**White**
1 B2	*2* F6
3 I5	*4* G8
5 H8	*6* H7
7 J6	*8* I7
9 J7	*10* I8
11 J8	*12* I10
13 F8	*14* E8
15 H10	*16* I9
17 G6	*18* G2
19 G5	*20* I2
21 H2	*22* H3
23 F4	

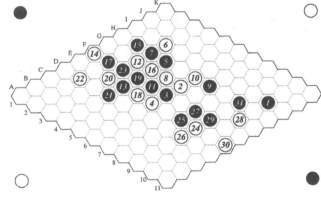

Sample Game 5:

Black	**White**
1 I10	*2* G6
3 F6	*4* E6
5 H4	*6* I3
7 H3	*8* G5
9 H7	*10* H6
11 F5	*12* G3
13 E4	*14* F1
15 H2	*16* G4
17 F2	*18* E5
19 F4	*20* E3
21 D4	*22* D2
23 F3	*24* E9
25 E8	*26* D9
27 F8	*28* G10
29 F9	*30* E11
31 H9	

Sample Game 6:

Black	White
1 K10	**2** F6
3 C7	**4** D4
5 E5	**6** I9
7 D7	**8** D8
9 B9	**10** B8
11 C8	**12** B10
13 F8	**14** F3
15 H4	**16** G4
17 G3	**18** F4
19 C4	**20** D3
21 B3	**22** C2
23 B2	**24** C3
25 H3	**26** G7
27 I6	**28** H8
29 G6	**30** F7
31 G5	**32** F5

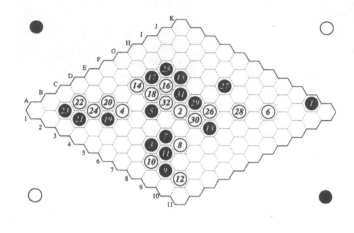

Sample Game 7:

Black	White
1 B3	**2** F6
3 D7	**4** D6
5 H4	**6** I2
7 C7	**8** C5
9 E6	**10** F4
11 D5	**12** D8
13 B9	**14** B10
15 C9	**16** C10
17 D9	**18** D11
19 E10	**20** E11
21 G10	**22** F10
23 G9	**24** F9
25 G8	**26** F3
27 E3	**28** F2
29 F8	**30** E9
31 E8	

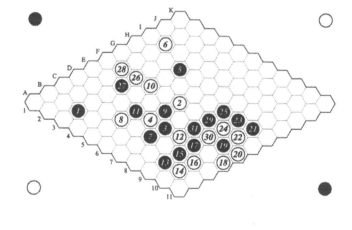

Sample Game 8:

Black	White
1 I2	*2* F6
3 C7	*4* C9
5 H8	*6* G9
7 G8	*8* I8
9 I5	*10* I4
11 G5	*12* G6
13 D8	*14* D7
15 C8	*16* E4
17 D4	*18* D10
19 E9	*20* D5
21 B6	

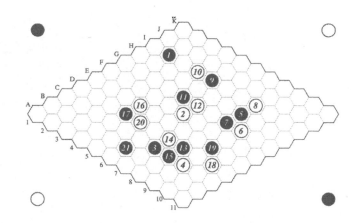

Sample Game 9:

Black	White
1 C2	*2* E6
3 F6	*4* E9
5 G8	*6* H4
7 G4	*8* F7
9 H6	*10* G6
11 D7	*12* E7
13 I4	*14* I3
15 K2	*16* J3
17 K3	*18* J4
19 K4	*20* J5
21 K5	*22* I7
23 K7	*24* J7
25 B8	*26* D6
27 A7	*28* B9

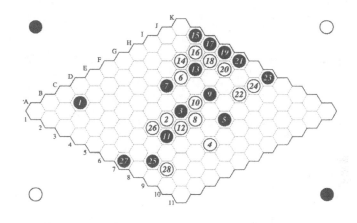

Sample Game 10:

Black	White
1 K1	*2* F6
3 I5	*4* H8
5 G7	*6* F9
7 H9	*8* E8

Black	White (Sample Game 10)
9 J7	*10* I6
11 E7	*12* C9
13 C8	*14* B8
15 D8	*16* D9
17 K5	*18* J3
19 J5	*20* H4
21 G5	*22* F5
23 E5	*24* H5
25 H6	*26* F7

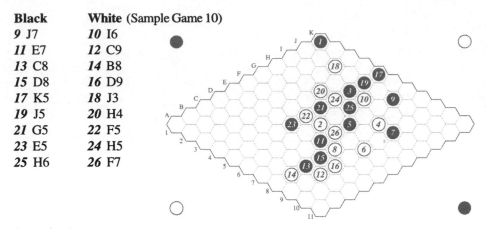

Sample Game 11:

Black	White
1 A2	*2* F6
3 C7	*4* D5
5 E6	*6* F4
7 G5	*8* C9
9 D9	*10* I9
11 I8	*12* H3
13 I3	*14* J2
15 B8	*16* H6
17 E4	*18* E5
19 C5	*20* C10
21 D8	*22* D10
23 E9	*24* E10

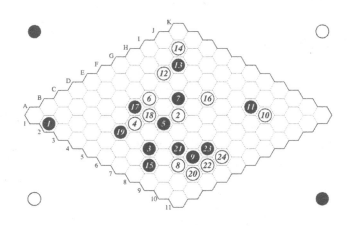

Sample Game 12:

Black	White
1 B2	*2* F6
3 C7	*4* D5
5 C5	*6* C6
7 A7	*8* B4
9 B6	*10* D3
11 F4	*12* E5
13 D4	*14* E3
15 E4	*16* G2
17 H3	*18* B9

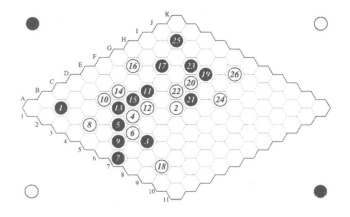

Black   **White** (Sample Game 12)

19 I5	*20* H5
21 G6	*22* G5
23 I4	*24* H7
25 J2	*26* J6

Sample Game 13:

Black   **White**

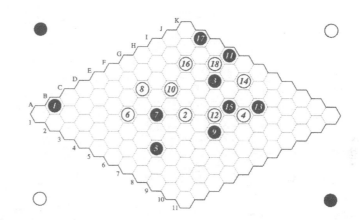

1 B1	*2* F6
3 I5	*4* H8
5 C7	*6* D4
7 E5	*8* F3
9 F8	*10* G4
11 K4	*12* G7
13 I8	*14* J6
15 H7	*16* I3
17 K2	*18* J4

Sample Game 14:

Black   **White**

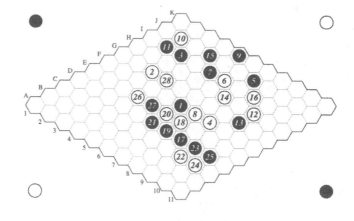

1 F6	*2* G3
3 I3	*4* F8
5 J7	*6* I6
7 I5	*8* F7
9 K5	*10* J2
11 I2	*12* H9
13 G9	*14* H7
15 J4	*16* I8
17 D8	*18* E7
19 D7	*20* E6
21 D6	*22* C9
23 D9	*24* C10
25 D10	*26* E4
27 E5	*28* G4

Sample Game 15:

Black	White
1 I10	*2* F6
3 H5	*4* H4
5 I3	*6* G7
7 F7	*8* H8
9 G8	*10* I4
11 E5	*12* F2
13 D7	*14* G3
15 F3	*16* G2
17 D3	*18* D6
19 E6	*20* C9
21 E8	

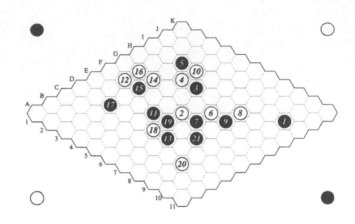

Sample Game 16:

Black	White
1 K10	*2* F6
3 I5	*4* H7
5 G7	*6* H5
7 G5	*8* G6
9 I4	*10* H6
11 I7	*12* I6
13 K5	*14* J2
15 J4	*16* H3
17 G3	*18* H4
19 C7	*20* D7
21 E6	*22* E7
23 C8	*24* D5
25 F4	*26* D6
27 C5	*28* C6
29 A7	*30* B10

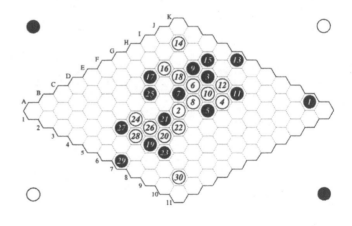

Sample Game 17:

Black	White
1 B1	*2* F6
3 C7	*4* D5
5 E5	*6* D7

Black   **White** (Sample Game 17)

Black	White
7 B6	*8* C8
9 C9	*10* B9
11 E7	*12* E6
13 G6	*14* G5
15 H5	*16* H4
17 I4	*18* I3
19 K2	*20* J3
21 K3	*22* J4
23 K4	*24* J5
25 K5	*26* H9
27 J7	*28* I8
29 I10	*30* J8
31 I7	*32* D9
33 G8	

Sample Game 18:

Black   **White**

Black	White
1 C3	*2* F6
3 D7	*4* D8
5 E8	*6* E7
7 B9	*8* B10
9 C9	*10* B8
11 C8	*12* D4
13 E5	*14* F3
15 G4	*16* F10
17 G9	*18* H2
19 I2	*20* H3
21 I3	*22* H4
23 I4	*24* H5
25 I5	*26* H6
27 I6	*28* H7
29 I7	*30* I9
31 J9	*32* I10

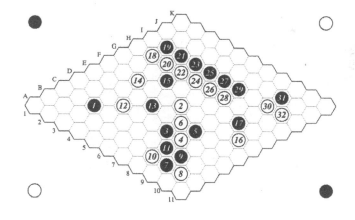

Sample Game 19:

Black	White
1 C3	*2* F6
3 I5	*4* G8
5 H8	*6* I6
7 H6	*8* H7
9 F7	*10* G6
11 G7	*12* E9
13 D9	*14* E8
15 F9	*16* F8
17 D8	*18* E7
19 D7	*20* E3
21 E5	*22* E4
23 C5	*24* D6
25 E6	*26* B7
27 F4	

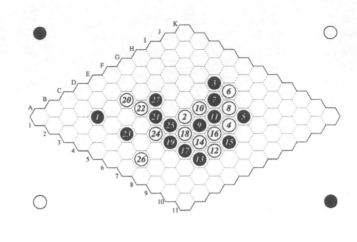

Sample Game 20:

Black	White
1 C6	*2* E7
3 D8	*4* C8
5 D7	*6* D9
7 D10	*8* D4
9 E4	*10* E6
11 D6	*12* E5
13 D5	*14* C9
15 F8	*16* E8
17 H5	*18* H4
19 J2	*20* G6
21 I4	*22* H7
23 H6	*24* G7
25 K6	*26* I7

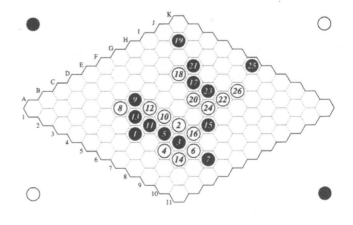

Sample Game 21:

Black	White
1 K3	*2* G6
3 D7	*4* E5
5 F5	*6* E6
7 F6	*8* E8
9 C9	*10* E7
11 H7	*12* I6
13 F7	*14* G3
15 H4	*16* F8
17 G7	*18* J1
19 J3	*20* G4
21 G5	*22* I3
23 I4	*24* K1
25 D4	*26* E3
27 D3	*28* E2
29 D2	*30* C6
31 F4	*32* H8
33 G8	*34* F10
35 I9	*36* G11
37 I10	*38* G9
39 J7	

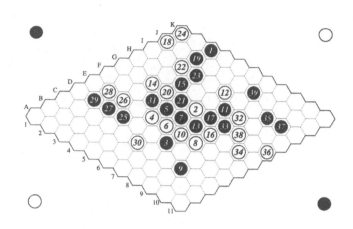

Sample Game 22:

Black	White
1 B3	*2* F6
3 D7	*4* D6
5 E6	*6* D8
7 C9	*8* C8
9 E7	*10* E8
11 G7	*12* H5
13 F7	*14* F9
15 H8	*16* F4
17 I3	*18* H7
19 J5	*20* H6
21 I7	*22* J4
23 G8	*24* F3
25 G2	*26* F5
27 C5	*28* D4

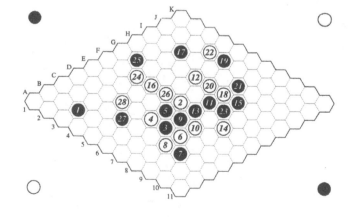

Sample Game 23:

Black	White
1 B2	*2* C4
3 H7	*4* G9
5 F9	*6* F8
7 H10	*8* I6
9 H6	*10* H5
11 F5	*12* G6
13 I5	*14* I8
15 G8	*16* E11
17 J9	*18* F10
19 K7	*20* K6
21 J7	*22* I7
23 J6	*24* J5
25 K5	*26* K4
27 M3	*28* K8
29 L7	*30* I10
31 J10	*32* L4
33 M4	*34* J8
35 C12	*36* D10
37 L8	*38* J11
39 K10	*40* M2
41 L3	*42* L9

Black	White
43 K9	*44* I13
45 G12	*46* H11
47 I11	*48* H12
49 I12	

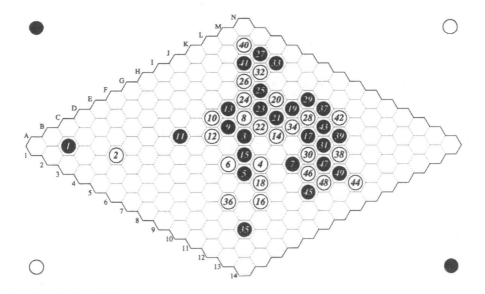

Sample Game 24:

Black	White
1 A2	*2* H7
3 E8	*4* F5
5 C6	*6* C7
7 B7	*8* E10
9 G9	*10* F8
11 E9	*12* C11
13 G7	*14* G8
15 F9	*16* F6
17 F7	*18* H5
19 I5	*20* H6
21 I7	*22* I6
23 K5	*24* H10
25 J7	*26* I9
27 F11	*28* H8
29 J9	*30* J8
31 L7	*32* I11
33 K10	*34* K9
35 E6	*36* E7
37 M8	*38* K6
39 M5	*40* L11

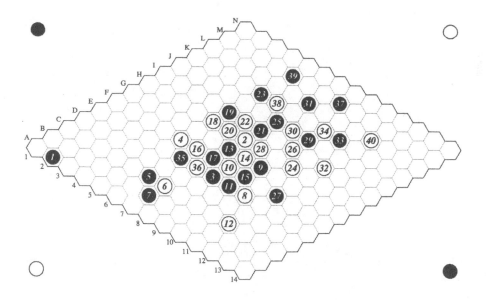

Sample Game 25:

Black	White		Black	White
1 B13	*2* G8		*41* F6	*42* G6
3 F8	*4* F4		*43* E7	*44* E8
5 G4	*6* G5		*45* D8	*46* C11
7 H5	*8* G7		*47* C10	*48* C9
9 J7	*10* K6		*49* B10	*50* B12
11 I6	*12* I5		*51* D10	*52* D12
13 H6	*14* J6		*53* D11	*54* C12
15 H8	*16* I7		*55* L6	*56* L5
17 H7	*18* I3		*57* M5	*58* L4
19 K3	*20* J4		*59* F11	*60* D9
21 J5	*22* K4		*61* B9	*62* C2
23 J3	*24* I4		*63* A3	*64* B1
25 H4	*26* H2		*65* D2	*66* C8
27 D6	*28* E3			
29 D3	*30* D5			
31 G3	*32* G2			
33 D4	*34* F3			
35 E5	*36* D7			
37 E4	*38* F5			
39 E6	*40* F7			

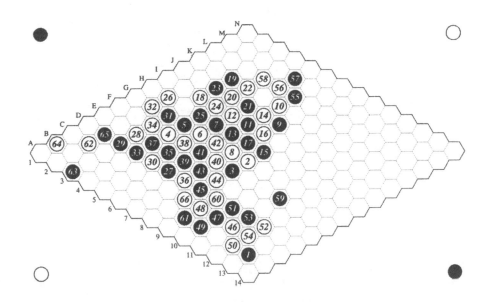

Sample Game 26:

Black	White
1 L13	*2* H7
3 E8	*4* E7
5 F7	*6* G5
7 F5	*8* F6
9 H6	*10* I5
11 H5	*12* I3
13 K3	*14* K2
15 J2	*16* K6
17 H4	*18* D3
19 F9	*20* H3
21 G4	*22* G3
23 F3	*24* F4
25 I4	*26* J3
27 N2	*28* L4
29 M4	*30* L7
31 C5	*32* E4
33 C3	*34* B7

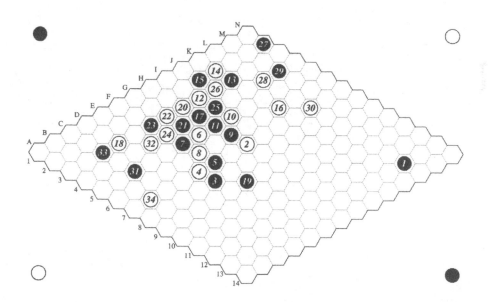

Sample Game 27:

Black	White
1 A14	*2* H7
3 E8	*4* D9
5 D8	*6* E9
7 B10	*8* D6
9 B9	*10* G5
11 E6	*12* F4
13 D5	*14* E5
15 C6	*16* C4
17 H6	*18* I4
19 J4	*20* I5
21 J5	*22* I6
23 K7	*24* J10
25 H9	*26* J7
27 L7	*28* I9

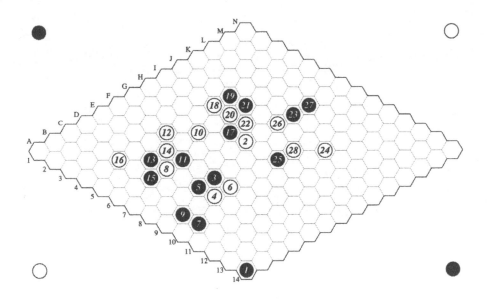

Proofs

This section gives informal outlines of some of the more interesting mathematical proofs regarding Hex. For more rigorous treatments of each proof, the reader should consult the references cited.

D.1 One Player Must Win

A game of Hex cannot end in a draw: one player must win. To prove this, it is sufficient to prove that a board completely filled with Black and White pieces must contain a winning path for either player. The following outline closely follows the proof described in [Binmore 1992] and [Van Rijswijck 1998].

Consider the following filled board:

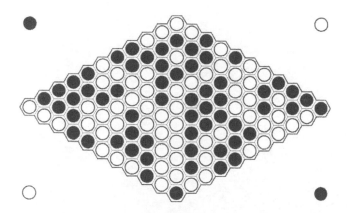

Let us now denote the edges belonging to each player by a row of cells occupied by pieces of that player's color, in the style of Piet Hein's original board design (see Section 1.7.2).

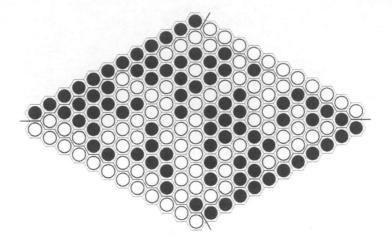

It is possible to mark four tags on the board, one at each corner, corresponding to edges within the border layer that divide pieces of opposite color.

Now starting at one of these tagged edges, for instance the leftmost one, a path is traced as follows:

• *at each intersection, turn either left or right such that the path continues to run between pieces of opposite color.*

This determines a unique path that is guaranteed not to visit any vertex twice. The path therefore cannot terminate on the board, and in fact can only terminate at one of the tagged corner edges other than that from which it began [Binmore 1992].

By the fact that the path is connected and must join two corner points, the path must therefore describe a connected link for one player between their edges and, therefore, the win.

The path traced from the leftmost corner tag terminates at the top corner in the above example and is a win for White. The complete path is shown below:

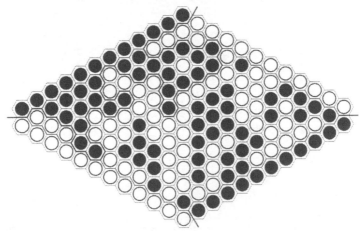

Van Rijswijck points out that the fact that a completely filled Hex board must contain a winning chain is equivalent to Brouwer's Fixed-Point Theorem. See [Gale 1979].

A similar proof is described in [Pierce 1980]. More inductive approaches to proving this proposition may be found in [Beck 1969] and [Berman 1976]. A pair of straightforward approaches that are not quite proofs are described in [Johnson 1976] and [Johnson and Zwillinger 1976].

D.2 First Player to Win

Given the fact that either White or Black must win the game, it's now possible to show that the first player has a winning strategy for any *nxn* board. Martin Gardner provides the following *existence proof* based on Nash's original *strategy stealing* argument [1959] that the first player has a theoretical win:

I   Either the first or second player must win; therefore, there must be a winning strategy for either the first or second player.

II   Assume that the second player has a winning strategy.

III The first player opens with an arbitrary move, then on subsequent moves adopts the second player's winning strategy. If on playing the winning strategy the first player is forced to move in a position that is already occupied by their last arbitrary move, then they make another arbitrary move elsewhere on the board.

IV   The extra piece belonging to the first player cannot harm their position: an extra piece is always an asset in Hex. Therefore the first player has a winning strategy.

V   This contradicts assumption **II** that the second player has a wining strategy.

VI   Therefore the first player has a winning strategy.

Note that this argument only proves that a winning strategy exists for the first player; it gives no hint as to what the actual strategy is. As Beasley puts it, this is a case of "when you know who, but not how" [1989].

Variations on this proof are discussed in a number of sources including [Beck 1969], [Pierce 1980], [Berlekamp et al. 1982], [Beasley 1989], [Binmore 1992], and [Van Rijswijck 1998]. It is equally valid for any game to which conditions **I** and **IV** apply.

D.3 Acute Corner is a Losing Opening

The two acute corner hexagons A1 and K11 are losing openings [Beck 1969]. If White plays *1* A1 then Black's reply *2* A2 essentially removes this piece from the game. The set of empty points touching *1* is reduced to the single point *t*, and if White's piece *1* were to play any further part in the game, it would have to connect through this point *t*. However, such a connection is superfluous as any move at point *t* connects White to their edge anyway. Black can pretend that their reply *2* is effectively the first move of the game, and wrest the first move advantage from White. After *2* A2, White has gained two points of territory *t* and *u*, whereas Black has gained no territory whatsoever.

This exact situation is used by Beck to prove that the second player has a winning line in Beck's Hex. This is a version of Hex in which the second player dictates where their opponent must play the opening piece in order to reduce the first move advantage. If they force their opponent to play in the acute corner, they should win with correct play.

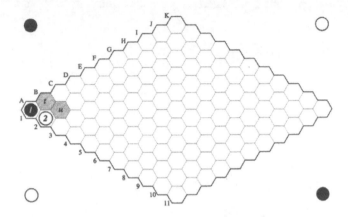

See Section 9.1.3 for a discussion of this point in relation to the swap option.

D.4 First Player Loses on $n*(n + 1)$ Board Playing Wide

Boards with $m * n$ sides (where $m \neq n$) are sometimes used in handicap play to equalize the game. However, if the first player must connect the more distant sides, as is usually the case, the second player can always win on an $n * (n + 1)$ board. Martin Gardner describes the following pairing strategy [1959], which is also discussed in [Binmore 1992].

Black (playing wide) moves first on the 6x7 board shown. White's strategy at each turn is to occupy the cell with the same label as the cell in which Black last moved.

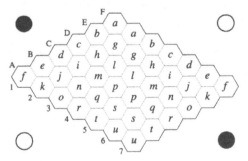

D.5 No Simultaneously Opposed 0-Connected Spanning Paths

It is not possible for both players to have 0-connected spanning paths at the same time. In fact, if one player has a 0-connected spanning path, then the best spanning path their opponent can possibly have is 2-connected. A brief proof can be outlined as follows:

A 0-connected spanning path cannot be beaten and provides a win for the player. This is due to the following property of 0-paths described in Section 4.3.5:

I   Each 0-path must contain at least one dual of empty points between its source and target groups.

This is a requirement of the rules of path consolidation. If there is not at least one dual of empty points within the path, then the pieces involved either form a single chain or a path that is not 0-connected.

The general case of a 0-path between two White pieces is illustrated below.

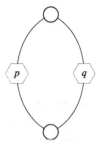

Points p and q are a pair of dual connection points within the 0-path. The path must contain at least one such set of dual points otherwise it is fully connected by adjacent moves and collapses into a single chain. Each solid arc represents a 0-connected path that may itself be composed of nested dual pairs forming 0-connected subpaths.

For the case of spanning paths, the two White end pieces represent the board edges. This path must be crossed if Black is to achieve a spanning path, as it is not possible to play outside the board and around either of White's end pieces.

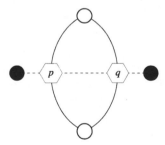

Black's spanning path passes through White's dual points p and q. However, these points lie in series along Black's path making it a 2-path.

Recall from the definition of path consolidation in Section 4.3.4 that:

I   2-paths cannot be safely consolidated.

This means that a 0-connected path can contain no 2-connected subpaths. The rationale for this constraint can be explained simply: if the opponent intrudes into a 2-path then the player requires two replies to defend the connection; however, they only have one chance to reply before the opponent can again intrude on their next turn.

To achieve a 0-connected spanning path, Black must use an intermediate piece to form duals of the two pairs of empty points shared with White's path as follows:

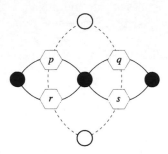

However, this means that White's spanning path now contains empty points in series without duals (which constitute 2-connected components) and therefore cannot be consolidated to a 0-connected spanning path for White.

Hence, the existence of a 0-connected spanning path for one player precludes the existence of a 0-connected spanning path for their opponent. Further, the connectivity of the opponent's best spanning path can be no better than 2 due to **II** above.

Hex Variants

This section contains a list of the more interesting connectivity games with some relationship to Hex. David Parlett places Hex in context of the overall history of board games as belonging to the class of *Games of Linear Connection*, within the more general category of *Space Games* [1999]. Connectivity games are distinct from other types of abstract board games with superficial similarities, such as Noughts and Crosses or Go-Moku, in that their winning conditions are based on the connection between target areas rather than patterns formed by the piece placement itself.

Only games invented since the release of Hex in the 1940s are included. Go, for instance, has many similarities but could not be considered a Hex variant. The inclusion of games in the following list does not imply that they were necessarily derived from or inspired by Hex. The list is not exhaustive, and it's regrettable that some worthy games may have been overlooked.

Alternative board representations that do not affect the way the game is played (such as those discussed in Section 1.7) are simply board modifications rather than Hex variants. Only two-player games are considered for reasons given in Sections 1.5.7 and 2.3.5.

E.1 Variants on the Hex Board

The following games can be played on the standard Hex board (or any *nxn* Hex board). Only the rules or objectives have been changed.

E.1.1 Beck's Hex

Other Names: Bex.

Origin: Anatole Beck [1969].

Rules: The same as for Hex except that the second player dictates where the first player moves.

Analysis: Second player can achieve a winning advantage by forcing the first player to open in one of the acute corners A1 or K11. Hex with the swap option is effectively a fairer version of Beck's Hex.

E.1.2 Kriegspiel Hex

Origin: William McWorter [1979].

Rules: As for Hex except that neither player has a view of the board. This requires the presence of an umpire to play the moves on the board and inform players when it is their turn, declare whether a move is illegal, and declare when a game has been won.

Analysis: As players do not know their opponent's moves except by encountering occupied points on a trial-and-error basis, Kriegspiel Hex is no longer a deterministic perfect information game. Some simple problems and proofs are discussed in [Broline 1981].

E.1.3 Reverse Hex

Other Names: Misere, Rex.

Origin: Ronald Evans [1974].

Rules: As for Hex except that the first player to complete a connection between their sides *loses* the game.

Analysis: Evans proves that White has a winning strategy with first move on even-sided boards.

E.1.4 Vex

Origin: Ronald Evans [1975-76] (named by David Silverman).

Rules: First player opens in an acute corner, and wins if they form a connected chain from this piece to either of the opposite edges. For instance, if White opens at A1, they must connect to either the top right or bottom right edges (or both!) to win.

Analysis: Variations on this game include Reverse Vex (first player loses if they are forced to complete a connection to either opposite edge), Vertical Vex (first player opens on an edge hexagon and wins if they connect this piece to the opposite edge), and Reverse Vertical Vex (first player opens on an edge hexagon and loses if they are forced to connect this piece to the opposite edge). These games are described as vertical due to the board representation used by Evans; the board is oriented so that two edges lie horizontal.

E.1.5 Handicap Hex I

Origin: Traditional.

Rules: As for Hex except that the weaker player starts the game with additional pieces on the board.

Analysis: If players are closely matched, then the stronger player places the extra pieces at hexagons of their choice; otherwise, the weaker player gets to place the pieces. This is a dangerous game to play—a match on the 11x11 board can be won with three well-placed pieces before the opponent has moved!

E.1.6 Handicap Hex II

Other Names: Uneven Hex, Asymmetrical Hex.

Origin: Traditional.

Rules: Play occurs on an *mxn* board where *m≠n*. The weaker player's direction of play is between the less distant edges (the smaller of *m* and *n*).

Analysis: There exists an explicit winning strategy for the weaker player under some circumstances (see Proof D.4). It is recommended that this method of handicap be used only if the weaker opponent has a very poor understanding of the game.

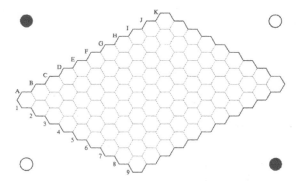

Figure E.1. A handicap 11x9 board. The weaker player takes Black.

In general, the handicap forms of Hex lead to unstable and unsatisfactory games, and should only be used if the opponents are wildly mismatched. If both players are of approximately equal strength, it is recommended that the swap option be used to equalize the first move advantage. If players are moderately unevenly matched then it is often sufficient to allow the weaker player to move first without threat of swap so that they may occupy the strong central hexagon F6.

Handicap II can be played on a standard board by deeming one of the weaker player's edges to be moved in a certain number of rows. Those rows are deemed out of bounds and not used during the game.

E.1.7 Modulo Hex

Other Names: Mod Hex.

Origin: Original.

Rules: As for Hex except that the board wraps or repeats periodically across its edges (an nxn board repeats itself modulus n across its edge). The edges do not form boundaries and players must align pieces across the edge to form a winning chain.

Analysis: Play occurs on a standard Hex board, which can be considered to repeat periodically in four directions away from each edge. For instance, the 7x7 game shown in Figure E.2 is a win for Black, as can be seen when copies of the board are aligned with each edge as shown in Figure E.3. It is possible for Black to trace a connected path that crosses both the top left and the bottom right edges of the original board. Ties may occur in Mod Hex.

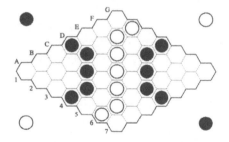

Figure E.2. A game of 7x7 Mod Hex won by Black.

Section 2.3.4 discusses how Modulo Hex can be projected onto the hemisphere, allowing a spherical or fisheye mapping that scrolls across the infinitely repeating board and shows all pieces at all times.

Variations on this game include:

•   *using the Surrounded Cell rule as an additional winning condition: a player wins if they complete a continuous chain that surrounds at least one cell (empty or occupied).*

•   *Antipodean goals: if the game is visualized on the sphere, the first player's goal is to form a continuous chain of pieces between the central hexagon of one board and the central hexagon of an adjacent repetition (odd-sided boards only).*

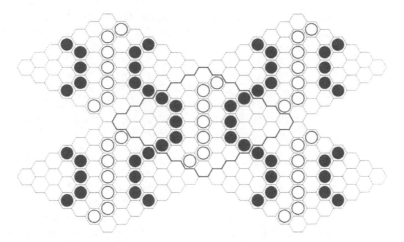

Figure E.3. Black's win is more obvious after placing copies of the board around each edge.

E.2 Other Hexagonal Connectivity Games

The following connectivity games are based on an underlying hexagonal structure and are closely related to Hex, but are not played on the Hex board.

E.2.1 Tex

Other Names: Infinite Hex.

Origin: Ronald Evans [1975-76].

Rules: Tex is played on an infinite tiling of hexagons (the name refers to the size of the board). The second player wins if they surround the opening player's piece with a connected chain of their pieces. An implied rule is that intermediate cells (empty or occupied) may also be included in the surrounded region. This is an example of the *Surrounded Cell* winning condition (see Section 2.3.4).

Analysis: Tex is quite different from Hex in that it is not constrained by edges which provide a reference point for each board cell, hence the need for a self-referential winning condition based on piece placement. Many Hex strategies such as edge templates and ladder escapes are of limited use in this case unless some pieces happen to form a continuous line that approximates an edge. This makes Tex easier to analyze in some respects.

Reverse Tex is not appropriate given the infinite nature of the board; the opponent can just reply to any move an infinite distance away, never threatening to form a connection. Edges are required to reduce the opponent's options and force losing moves in Reverse Tex.

Tex is similar in principle to Modulo Hex but is truly infinite and unconstrained by periodic repetition.

A fisheye lens scrolling over the infinite playing surface may provide a suitable method of representing the large board to the player, but has the disadvantage that only a finite portion of the board can be seen at any one time.

E.2.2   Y

Origin: Schensted and Titus describe the invention of Y in 1953 in their book *Mudcrack Y & Poly-Y* [1975], reviewed in [Mallett 1998].

Rules: First player to form a connected chain linking all three sides wins. Each side is shared by both players.

Analysis: Y is probably the non-Hex variant that bears the closest relationship to Hex. It cannot end in a draw. Most Hex strategies can be successfully applied. Edge ladder templates are still relevant, but overlap can occur between both players' templates to the same edge without invalidating them, as each player shares each edge. Y is discussed in detail in Section 2.3.2.

The commercial version of Y is played on a board containing two or three pentagons placed symmetrically near its center, and slightly curved around the edges. Schensted and Titus chose this design to reduce the playing distance between the central hexagon and the acute corners. Schensted found in 1969 that distorting the board regions greatly enhanced the importance of the corners [Schmittberger 1983].

Half-piece handicaps are allowed in Y, but are inappropriate for Hex. There exists a version called Mudcrack Y due to its irregular tiling, reminiscent of cracked mud.

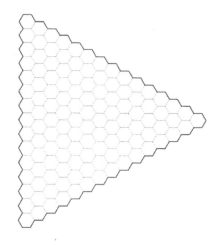

Figure E.4. The Y board (not the commercial version).

E.2.3 Poly-Y

Origin: Titus and Schensted [1975] but developed earlier.

Rules: Players capture corners on the board by establishing connected chains of pieces that link both edges adjacent to the corner, and one other edge on the board. The player who captures the most corners wins. Played on an odd-sided polygonal board with five or more edges. The board is typically tiled with hexagons, with additional polygonal tiles as required to complete the design.

Analysis: Poly-Y shares many aspects of Y but has more strategic depth. It is one of the more interesting of the Hex group of games.

E.2.4 Star

Origin: Craige Schensted, described in [Schmittberger 1983].

Rules: Players alternate placing pieces on unoccupied white hexagons of the board shown in Figure E.5 until both players choose to pass. Players score points for each connected chain of pieces that connect three or more shaded hexagons (these formations are called *stars* due to the typical shape that forms during play). Each star is worth 2 points less the number of shaded hexagons it touches. The player with the most points wins.

Analysis: Star evolved from Poly-Y and is one of the most difficult of the hexagonal connectivity games. Its invention was celebrated with widespread acclaim. Star has many interesting features, including the remarkable fact that the total number of points scored by both players is always two less than the number of shaded hexagons, regardless of board size [Schmittberger 1983]. Due to this, draws are not possible in Star.

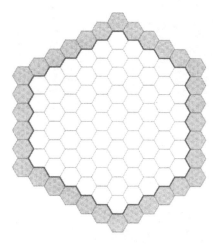

Figure E.5. The Star board.

E.2.5 SuperStar

Origin: Christian Freeling [1999].

Rules: Rules are similar to Star except for the following additions:

•   A *star* is a chain touching at least 3 cells of the edge. The value of a star is two less than the number of cells of the edge it touches.

•   A *superstar* is a chain connecting at least 3 sides. The value of a superstar is 5 * (S - 2), where S is the number of sides it connects.

•   A *ring* is a chain surrounding at least one cell. The value of a ring is one point for every vacant cell that it surrounds, and five points for each opponent's stone trapped within it.

Analysis: Superstar was developed from Star by adding a more complex playing area and additional scoring conditions to increase the complexity of the game.

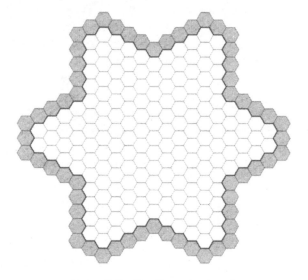

Figure E.6.   The Superstar board.

E.2.6 Metahex

Origin: Original.

Rules: As for Hex: players attempt to link their edges with a continuous chain of pieces.

Analysis: The Metahex is a hexagonally shaped board composed of hexagons. Metahex describes a board and a system of games that may be played upon it, rather than a particular game itself. For instance, each player may be allotted two adjacent sides and one opposite side, or each player may own three non-adjacent sides. Ties are possible in the latter case.

The Metahex board has the advantage that acute corners are eliminated (as Schensted desired with Y). The discrepancy between center-to-corner and center-to-edge distance is minimized. For this reason the Metahex board is suitable for experimentation and analysis (see for instance Section 2.3.5).

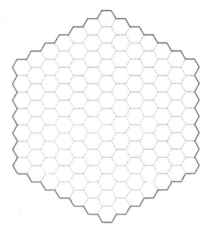

Figure E.7. The basic 7- Metahex board.

E.2.7 Havannah

Origin: Christian Freeling. Described in [Freeling 1999] but invented much earlier.

Rules: As for Hex except that there are three winning conditions. A player wins when they complete any of:

• a *ring* (chain around at least one cell),

• a *bridge* (chain linking two corners), or

• a *fork* (chain linking three sides—corners do not belong to sides).

Havannah is played on an *n*-Metahex board.

Analysis: Havannah contains many elements of Hex and Y but with added complexity through simultaneous winning conditions. This gives it somewhat less clarity than its precursors and makes it quite difficult to analyze. The *ring* winning condition is the same as the Surrounded Cell rule of Tex.

Players who enjoy Hex and Y will probably enjoy Havannah as an alternative approach to this class of games.

E.2.8 Cylindrical Hex

Other Names: Annular Hex.

Origin: Alpern and Beck [1991].

Rules: As for Hex, except that the board wraps or repeats periodically in one player's direction of play. The player in the wrapped direction must form a continuous chain of pieces that connects across the edge to win, whereas their opponent must simply connect the top and bottom edge of the cylinder.

Analysis: This game is similar to Modulo Hex except that the board repeats periodically only in one direction of play rather than both. This is equivalent to wrapping the game around a cylinder, which can then be squashed flat into an annulus for representational purposes.

    The process that converts rectangular Hex to Cylindrical Hex preserves winning paths for the player joining the ends of the cylinder, but not for the player wrapping around the cylinder. Alpern and Beck prove that Cylindrical Hex cannot end in a tie, and that the player playing end-to-end can always win if the board dimension in their direction of play is even.

    Figure E.8 shows a Hex board described as a skewed rectangular grid on the left (see Section 2.1). For the purposes of Cylindrical Hex, the board wraps in the horizontal direction. The cylindrical board mapped to the annulus is shown on the right.

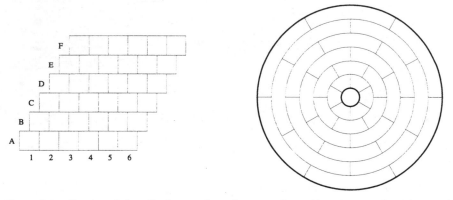

Figure E.8. Hex board described on a skewed rectangular grid and mapped to the annulus.

E.2.9 Hexbo

Origin: Created by Richard Rognlie, described in [Steere 1996].

Rules: The rules for Hexbo are not concise enough to describe here—refer to [Steere 1996]. The game is played on an *n*-Metahex board.

Analysis: An interesting game based on competition among tree roots growing within a hexagonal grid system. Hexbo was derived by Richard Rognlie from its rectangular grid equivalent Tanbo.

E.3 Non-Hexagonal Connectivity Games

The following games bear resemblance to Hex in their style of play or underlying principles of connectivity, but are not necessarily played on the hexagonal grid.

E.3.1 Gale

Other Names: Bridg-It, Birdcage, Connections, Connexxions.

Origin: Invented by David Gale, described in a number of sources including [Gardner 1966].

Rules: Play occurs on the edges of the board shown in Figure E.9. One player (Short) solidifies a dotted edge each turn, and their opponent (Cut) deletes a dotted edge each turn. Short wins if they connect the terminal nodes *s* and *s'* with a continuous chain of solid edges, otherwise Cut wins.

Analysis: Gale is a specialization of the Shannon Switching Game (as is Hex). For a more complete discussion of Gale, see Section 1.5.1.1. See also Connections [Parlett 1999] and Connexxions [Rognlie 1996] for descriptions of equivalent games.

The Gale board and a complete game are shown in Figure E.9.

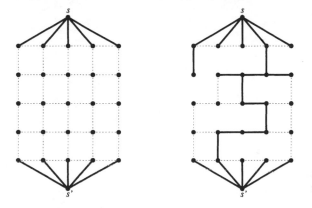

Figure E.9. The Gale board and a completed game.

E.3.2 TwixT

Origin: Invented by Alex Randolph, described in detail in [Wells 1973].

Rules: Similar to Gale, except that connections occur between pieces a knight's move apart rather than adjacent pieces. Connections are optional; players are not forced to solidify all possible connections resulting from a move. Play occurs on a 24x24 board.

Analysis: TwixT is a most interesting game. It has enjoyed greater commercial success than Hex and is quite appealing to play, perhaps because it is somewhat faster moving and has a greater feeling of space due to its non-adjacent connections.

TwixT has excellent clarity and depth of strategy on a par with Hex, and compares favorably with other connectivity game of this class.

E.3.3 Non-Regular Hex

Origin: David Book [1998].

Rules: As for Hex. Play occurs on a board composed of semi-regular or irregular tilings.

Analysis: Great variety in play can be achieved by playing Hex on tilings that are not hexagonal, nor even regular. Figure E.10 shows an irregular tiling that provides an interesting game. See Section 2.3.3 for a more in-depth discussion of this group of games.

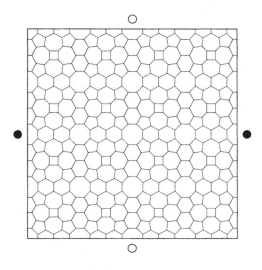

Figure E.10. An irregular tessellation that makes an interesting playing surface.

E.3.4 Map Hex

Origin: David Book [1998].

Rules: As for Hex. The playing surface is a tessellation defined by boundary marking on a map. Countries that share an edge (border) are deemed to be adjacent.

Analysis: Figure E.11 shows a game of Map Hex played on a map of the United States. Unfortunately, this map does not provide a very interesting game, and Black can win easily with first move. A tie can occur on this map, as four regions share a common vertex towards the lower left (David Book 1998).

Map Hex is a particularly appropriate form of the game that harks back to its origins— Hex originally occurred to Piet Hein while he was studying the famous four-color theorem of topology [Gardner 1959]. See Section 2.3.3.

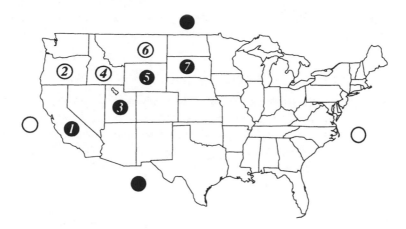

Figure E.11. A game of Map Hex won by Black (who can always win on this map with first move).

E.3.5 Voronoi Hex

Other Names: Vortex.

Origin: Original.

Rules: As for Hex. The game is played on a board defined by the Voronoi diagram of a randomly distributed set of points.

Analysis: The "dimensions" of the board are determined by the density of the point set. Care must be taken to ensure that the distribution is well-behaved; otherwise, the board may contain disproportionately weak or strong regions as shown in Figure E.12. The

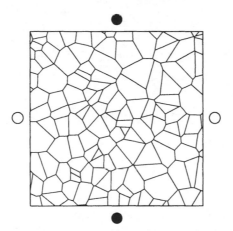

Figure E.12. A Vortex board. This point set is not particularly well-behaved.

optimal Vortex board contains mostly six-sided regions, but with enough random variation to make the game interesting.

E.4   Tile-Based Connectivity Games

The next set of games describe connections between adjacent pieces explicitly through the use of paths between tile edges. This allows a significantly different flavor of play within the overall context of connectivity games. The internal composition of pieces is no longer uniform, and inter-positional connections are subject to constraints in that each tile edge only connects with those other edges with which it shares a path.

E.4.1 Trax

Origin: Invented by David Smith, described in detail in [Bailey 1997].

Rules: Players take turns placing either of the tiles onto an unbounded playing surface such that the tile being placed touches at least one existing tile and path colors match across the edge to existing tiles. The first player to complete a loop of their color or a path of their color that spans eight columns or eight rows wins. Some additional restrictions regarding piece placement exist [Bailey 1999].

Analysis: Trax is a popular game, and deservedly so. It involves reasonably deep strategic play, and the development of paths during the course of a game results in patterns that are often aesthetically interesting.

Variations constrained to specific board sizes exist, for instance 8x8 Trax [Rognlie 1996].

Figure E.13. The two Trax tiles.

E.4.2 Thoughtwave

Origin: Invented by Eric Solomon, described in [Parlett 1999] and [Frogley et al. 1998].

Rules: Play occurs on a 10x10 square grid. Each player owns two opposite edges and starts the game with 24 of the tiles illustrated in Figure E.14 in the following quantities: 1:5:10:6:2. Players alternate placing a tile on the board until one player wins by completing a continuous path between their edges.

Analysis: The unique feature of this game that distinguishes it from other games in this class is that both players share all paths on the board. Any tile placed on the board may be used in either player's winning path.

It is conjectured that an interesting game may be devised by combining this approach of shared paths with the added complexity of the hexagonal grid. The author is currently experimenting with games of this nature.

Figure E.14. The five Thoughtwave tiles.

E.4.3 The Black Path Game

Other Names: A Winding Road, Black's Road Game, Snake's Road, Squiggly Road.

Origin: Invented by William Black, described in [Berlekamp et al 1982] and [Koch 1991].

Rules: Play occurs on an *mxn* rectangular grid, typically 8x8. The first player opens by placing one of the tiles shown in Figure E.15 on a corner square. Players then alternate placing tiles on any empty square such that the tile being placed becomes part of a continuous path whose source can be traced to the opening tile. The first player that is forced to terminate this opening path by playing it into an edge loses the game.

Analysis: This game is essentially a type of Reverse 8x8 Trax or Reverse Thoughtwave. The fact that the winning condition relies on manipulating the opponent's position makes this game's strategy somewhat less tangible and pleasing than other games in this class that emphasize development of the player's own position instead.

It is interesting to note that playing this game with the second and third tiles only leads to a very short game, but would result in a Truchet tiling [Pickover 1990].

Figure E.15. The three Black Path tiles.

E.4.4 Pipeline

Origin: Invented by Edward T. Okamura, described in [McCallion 1992].

Rules: Similar in principle to Thoughtwave but with some constraints added. Play occurs on a 15x15 grid with an expanded tile set of 200. The object of the game is to complete a pipeline from an oil well in the board's center to any of three loading docks at the periphery of the board.

Analysis: Pipeline is an attractively presented commercial game that presents a clever take on the connection theme. Most players will be familiar with the concept of extending pipelines and the need to keep them continuous; therefore, the game will feel more intuitive and "natural" than strictly abstract counterparts such as Hex and TwixT.

The game is made more interesting through the presence of eight obstacle squares on the board that must be circumnavigated. At the start of each round, players select five pipeline tiles from the stockpile to be used that turn.

E.4.5 Octiles

Origin: Invented by Dale Walton, described in [Thomashow 1986].

Rules: The Octiles set includes (among other items) eighteen octagonal tiles, each with a unique combination of four paths of the same color joining pairs of edges, such that every edge is linked to one other across the tile. Rules for three styles of game are provided, but are not concise enough to describe here in detail.

Analysis: Octiles is highly regarded and provides tremendous scope for variety; it's more of a puzzle kit than a single game. The rules booklet comes complete with several Octiles puzzles based on various configurations of pieces.

E.4.6 Kaliko

Other Names: Psyche Paths.

Origin: Invented by Craige Schensted and Charles Titus, described in [O'Sullivan 1998].

Rules: The Kaliko set consists of eighty-five hexagonal tiles with unique combinations of paths in three colors connecting pairs of edges, such that each edge maps to one other edge. Players start with seven tiles each, hidden from their opponents' view behind a screen. The game begins with a single tile in the center of the playing area. Players take turns adding tiles to develop connections from tiles already played, such that path colors match across edges and specific conditions regarding path continuity are met.

Analysis: This tile-based connection game is distinct in that it mirrors Scrabble in several aspects. Players may discard and replace unwanted tiles in lieu of playing their turn, and the scoring system used is reminiscent of that used in Scrabble.

E.5   Mobile Pieces

Several connectivity games based on the hexagonal grid allow the movement of pieces after they are played. Some of these games bear similarities with Hex, but in general this group diverges from the basic principles of Hex and will not be studied in detail. For further discussions of these games see the references cited:

*   Trinidad [Tijs Krammer 1999b],

*   Link [How 1984],

- Tack [How 1984],

- Fan [How 1984],

- Hexxagon [Rognlie 1996],

- Hexadame [Freeling 1999],

- HexEmergo [Freeling 1999],

- Hexade [Freeling 1999],

- Macbeth [Freeling 1999].

E.6   Three-Dimensional Connectivity Games

The connectivity games considered so far in this section have all occurred on a flat playing surface, including Cylindrical Hex, which simply involves the theoretical mapping of a surface onto a cylinder, and the hemispherical projection suggested for viewing Modulo Hex.

The most obvious extension of planar games to three dimensions involves stacking successive playing surfaces, with connection between touching cells of adjacent layers. Each planar cell is therefore extruded into a *prism* whose cross-section is the shape of the original tile.

This approach is suitable for games based on the square grid, as the extrusion of a square produces a cube—a platonic solid that is symmetrical about all axes, allowing most game rules to be easily adapted to three-dimensional play. However, this will not work so well for games based on the hexagonal grid, which extrudes to a prism that has distinctly different properties in respect to different directions of play. For instance, moves between layers are perpendicular to the players' target directions, which means that play tends to occur on a single layer and the full potential of three-dimensional play is not realized.

One way to overcome this problem is to use non-prismatic polyhedral cells that exhibit similar behavior in all directions of play. The cuboctahedron [Coxeter 1973] produced by *cannonball stacking* is an obvious candidate. This structure is equivalent to the most stable packing of spheres in space, which can be seen as a natural three-dimensional extension of the hexagonal arrangement that describes the most stable packing of circles in the plane [Ghyka 1977]. This arrangement has the desirable property that it describes a hexagonal basis across each layer of the stack. However, the cuboctahedron has a total of fourteen faces (giving fourteen potential connections to adjacent cells), making the task of blocking a connection to/from the cell extremely difficult; there are a number of alternative paths available to the player who needs to step around a blocking piece.

This has serious ramifications for connectivity games of the Hex class that rely on the conflict of crossing paths within the plane for their strategic depth. When moved out of the plane, this conflict is largely resolved (players may step over or under blockages), and superficial rules must be introduced to make the game interesting and avoid its degeneration into a simple race. As demonstrated in Section 2.3, hexagonal tiling offering six potential connections per cell is optimal for similar games of connectivity.

Alternative space packings such as that provided by the fourteen-sided tetrakaidekahedron [Weyl 1952] suffer the same problems.

There are also the purely practical problems that face three-dimensional versions of any game, such as how to present an unambiguous view of the board to both players, and how to place pieces within central cells that may be surrounded by pieces.

Blank Hex Boards

Blank Hex boards are provided for the reader's use, ranging in size from 3x3 to 26x26. Players should make a copy of the relevant board, then may choose to play Hex as a pen-and-paper game or mount the board copy on cardboard and play with light and dark counters (Go stones make excellent pieces).

More ambitious players may wish to create a proper wooden playing board by marking the outlines of the hexagonal grid with a tool such as a chisel or soldering iron on an appropriately shaped piece of timber. Marbles make good playing pieces and are readily available, but require that a circular indentation be drilled at the center of each hexagon for placement of the pieces.

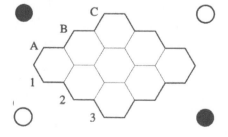

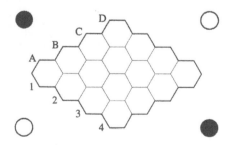

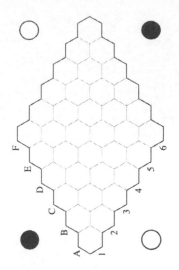

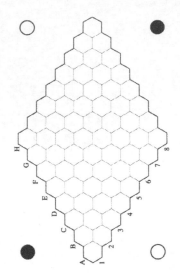

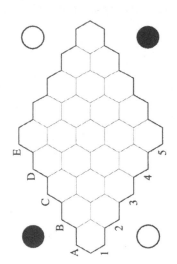

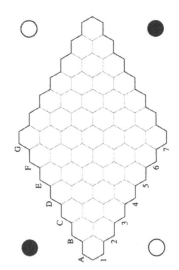

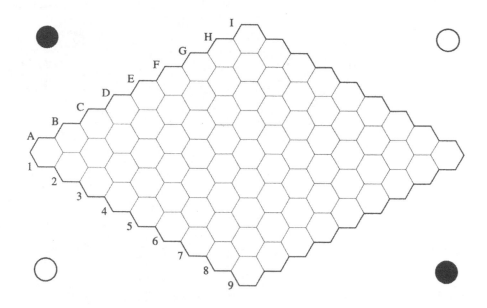

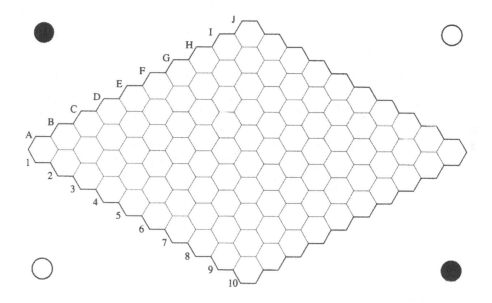

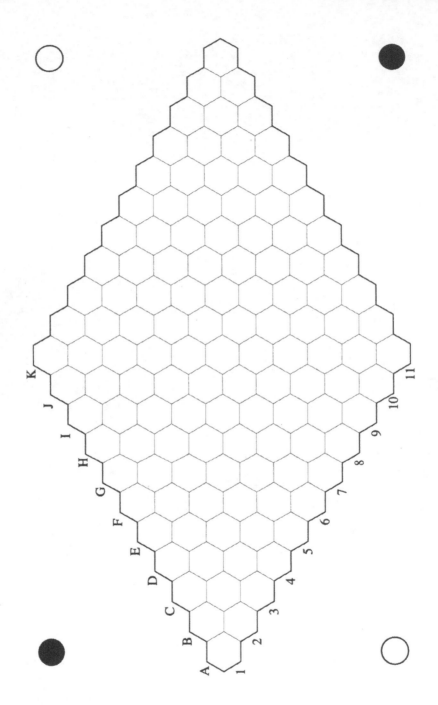

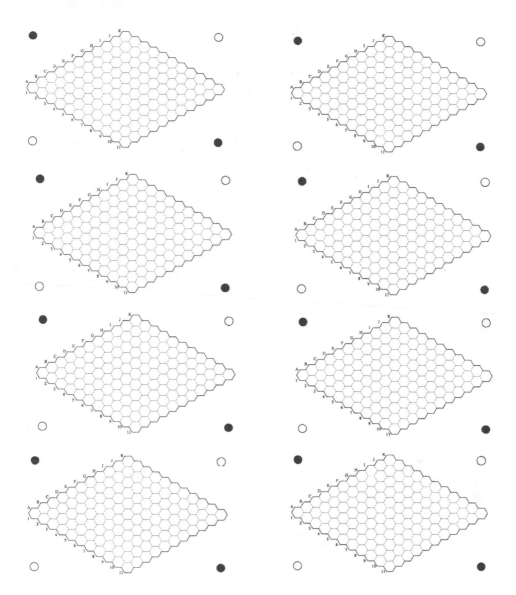

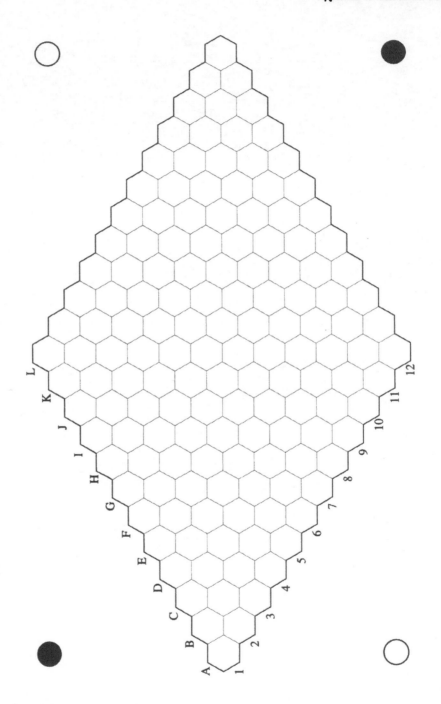

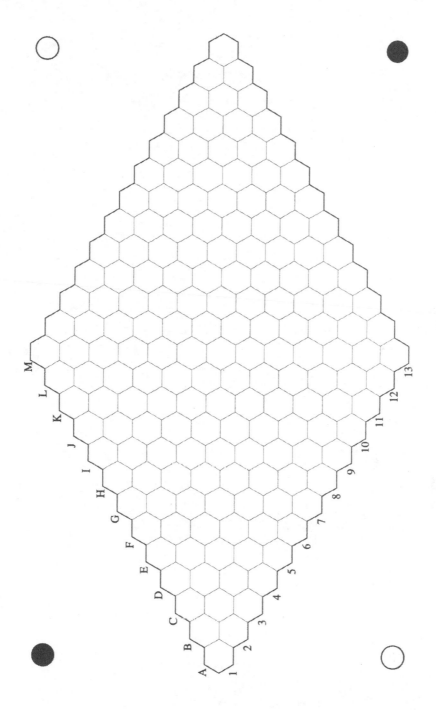

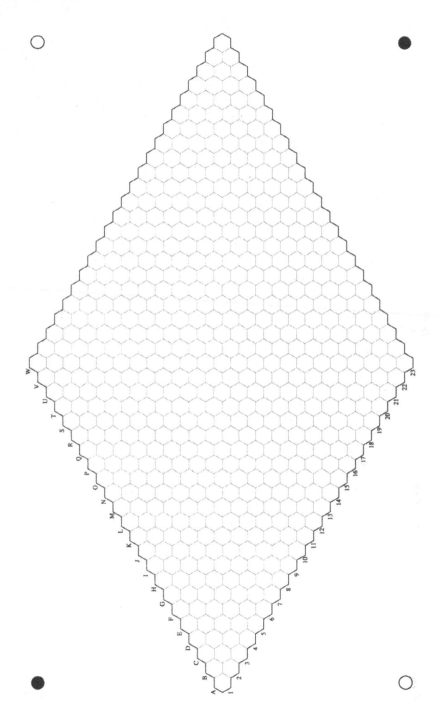

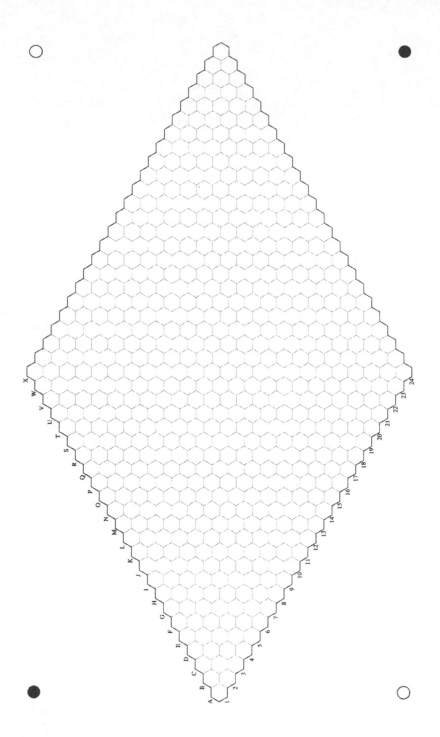

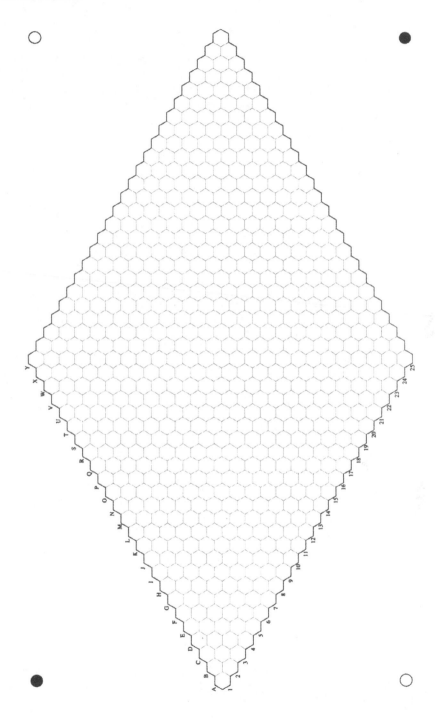

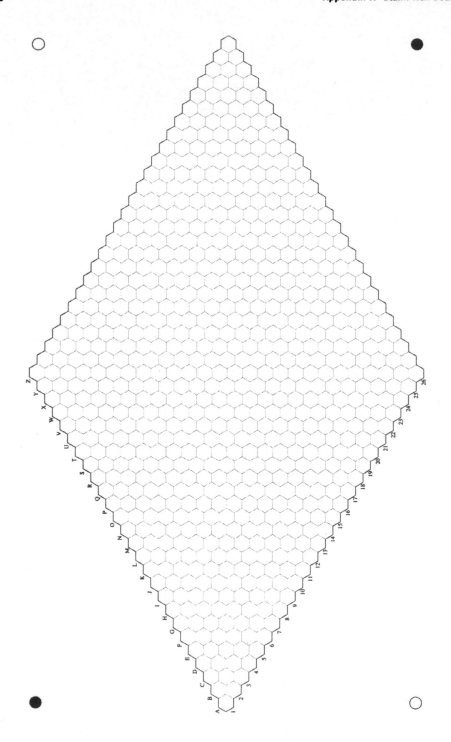

Polyhexes

A *polyhex* is a planar shape composed of connected equally-sized regular hexagons, such that any two hexagons either share exactly one edge or are disjoint [Tosic et al. 1995]. Chains of pieces formed during a game of Hex are polyhexes: For instance, consider the chains making up the following game:

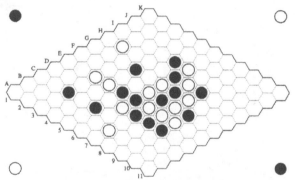

Figure G.1. A typical game of Hex.

These chains are equivalent to the five White and seven Black polyhexes shown in Figure G.2:

Figure G.2. White and Black polyhexes corresponding to the above chains.

It is convenient to describe polyhexes by their boundaries, although this means that shapes with interior holes are indistinguishable from those with identical boundaries but different or no interior holes. Boundary representation leads to compact coding schemes and efficient algorithms for classifying polyhexes. The exact encoding algorithm used by Tosin et al. is described in detail in their 1995 paper.

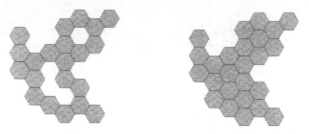

Figure G.3.   Polyhexes with identical boundaries.

Fortunately, this lack of interior knowledge is irrelevant to the description of piece chains in a game of Hex. Only exterior connections to other chains and edges are important; thus, existing polyhex enumeration algorithms can be used for describing Hex chains.

h	Total Polyhexes
1	1
2	3
3	10
4	32
5	96
6	272
7	736
8	1,920
9	4,864
10	12,032
11	29,184
12	69,632
13	163,840
14	380,928
15	876,544
16	1,990,656
17	5,421,984
18	10,158,080

Table G.1.   Total number of polyhexes from h hexagons (includes rotations).

This opens up intriguing possibilities for the efficient implementation of a Hex program. If it were possible to preclassify the various piece chain shapes and store them in a table by their unique code, information relevant to specific chains could be accessed with a quick lookup. This might include the list of steps from each chain (see Sections 4.2 and 8.1), connections between pairs of chain shapes, or other relevant details. Even allowing for the computational expense of locating and classifying each chain, this could provide considerable savings over generating the same information on-the-fly.

However, the complexity of the problem cannot be ignored. Tosic et al. use a brute force algorithm to enumerate the total number of unique polyhexes for each number of hexagons h from 1 to 18. From their results (shown in Table G.1), it can be seen that the number of combinations for polyhexes size 16 and greater becomes prohibitively large. But how often do chains of that size actually occur during play on an 11x11 board?

Table G.2 shows the number of occurrences of each chain size for a representative sample of 100 games played on the standard 11x11 board, taken at random from the database maintained on the Gamerz.NET server [Rognlie 1996]. Although complex chains of 15 and 20 pieces do occur, these are strictly in the minority and almost 99% of chains are eight pieces in size or smaller. This is encouraging as there are only 1,920 polyhexes (or unique

Chain Size	Per 100 Games
1	911
2	140
3	80
4	53
5	42
6	17
7	15
8	7
9	2
10	9
11	4
12	0
13	0
14	0
15	2
16	0
17	0
18	0
19	0
20	1

Table G.2. Distribution of chain sizes over 100 sample games.

chain configurations) with eight pieces or less. However, larger chains will obviously occur more frequently on larger boards.

Figure G.4 shows the first four sets of polyhexes without reflections or rotations. The set of 4-polyhexes is so distinctive that its members have been given common names, as shown in the diagram for convenient reference [Weisstein 1999].

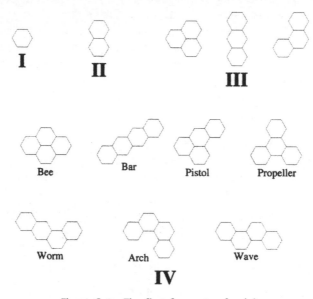

Figure G.4. The first four sets of polyhexes.

It makes sense to preclassify Hex chains up to a certain reasonable size, then do on-the-fly computation for larger chains if and when they arise. The threshold number would depend on the amount of information to be stored for each chain entry, but eight pieces appears to be a good estimate for the 11x11 board.

Larger polyhexes may be constructed from the combination of simpler ones; however, given that multiple sub-classifications and merging operations would be required, it may be more efficient to simply evaluate large shapes on-the-fly.

Hex Programs

Due to the complexity of the game, there are understandably few computer Hex programs available. Some commercial Hex programs exist, but these are generally not very impressive except in terms of design and packaging.

The best of the publicly-available programs capable of playing on boards up to 11x11 is *Hexy* by Vadim Anshelevich [1999], which can be downloaded from http://home.earthlink.net/~vanshel/. It is recommended that Hex players wishing to improve their game obtain a copy of *Hexy* for match practice and tutorial purposes.

Hexy plays a well rounded game, responds quickly, and has a simple but effective user interface. A technical paper describing Hexy's evaluation function can be found at [Anshelevich 2000].

Although an experienced player will beat *Hexy*, it offers a randomization feature so that the player is not subjected to the same sequence of moves every game. This addresses a serious shortcoming of many computer players in that once defeated, they can always be defeated by the same line of play.

Queenbee, a program developed by Jack Van Rijswijck, is reportedly a superior player but is not available for public use and has not been evaluated [Van Rijswijck 1999]. Hopefully this program will be made available shortly. A technical discussion of Queenbee can be found at [Van Rijswijck 2000].

Other players on the 11x11 board include Chris Lusby Taylor's program *Hex* developed in the early 1990s, which plays competently but suffers from some lack of strategy. This program can be downloaded from [Rusin 1998].

Hex-7 is an attractively presented program that plays a strong game interactively on a 7x7 board [MazeWorks 1999].

Glossary

This glossary defines terms, symbols and notation used in this publication. Refer to the index and text for more detailed explanations.

I.1 Terms

adjacent move foil   Ladder escape foil based on a piece adjacent to the escape piece.

adjacent   Two cells that share an edge are adjacent.

board position   A combination of Black and White pieces on the board.

bridge   A pair of non-adjacent pieces that share two adjacent empty points and can generally be considered to be safely connected.

cascading escape   A partial ladder escape that continues in the same direction as the original ladder.

chain   Maximal set of one or more pieces of the same color, such that every piece within the set can be reached from any other through a set of adjacent moves.

clarity   The amount of certainty with which a player can plan ahead and formulate strategies.

classic defense   The most natural position from which to block an advancing piece. This position is a bridge and adjacent step away from the advancing piece.

connectivity template   Predefined pattern of pieces that describes a safe connection.

dual   A pair of empty points within a path that allows safe consolidation to a 0-path. If the opponent intrudes in one of the dual points, the connection can be saved by playing in the other.

edge template   Connectivity template between a point and an edge, or a piece and an edge.

embedded piece   A piece belonging to the attacker surrounded by defender's pieces such that it has limited opportunities for connection. An embedded piece may lie adjacent to an advancing ladder yet not provide an escape.

empty point superset   A set of empty points belonging to a group, step, or path that is not the minimal set. Superfluous points should be discarded from the set.

escape piece   A piece belonging to the attacking player that enables a ladder escape. Escape pieces typically lie along the path of the advancing ladder or on a point nearby.

foldback escape   A partial ladder escape that doubles back in the direction opposite to that of the original ladder.

forced reply   Reply to a forcing move that a player is obliged to make.

forcing move   A move that poses a threat that cannot be ignored. Forcing moves can be used to maneuver the opponent into a disadvantageous position.

fork   A move that poses two or more immediate threats.

friendly intrusion   A move that intrudes into a player's own template, making it invalid. The template's connection is not threatened, but it can then be described more succinctly by two or more subtemplates incorporating the new piece.

goals   The two opposite edges of the board belonging to a player that they must attempt to connect.

group   A connected set of chains and empty points.

home area   The region of the board easily reached by either of the player's edges.

interior template	Connectivity template between two points, two pieces, or a point and a piece.
intrusion	A move that intrudes into an opponent's connectivity template. The opponent must answer the intrusion or the template is rendered invalid.
ladder escape foil	A sequence of moves by the defender that successfully blocks a ladder escape.
ladder escape fork	A killer move that provides both an escape for an advancing ladder (either formed or as yet unformed), and additional threat of connection that must be answered. Ladder escape forks are forcing moves.
ladder escape	A sequence of moves by which the attacking player is able to jump ahead of the ladder and complete their connection to the edge.
ladder	A sequence of pieces that progresses parallel to a player's edge. Ladders form as the attacking player attempts to connect to their edge while the defending player blocks the connection, resulting in a chain of pieces a constant distance from the edge.
loose connection	Two pieces one adjacent step removed from being a bridge. The positional relationship between the pieces is that same as that between the classic defense and an advancing piece.
momentum	The player with the initiative is said to have the momentum of the game. In a close game the momentum may swing with each move.
neighborhood value	Simple measure of a point's potential value in the game based on the arrangement of pieces in its neighborhood and its location on the board.
neighborhood	The set of six neighbors immediately adjacent to a point.
non-adjacent foil	Ladder escape foil based on a piece that is not adjacent to the escape piece. Typically this is only possible when the additional threat posed by an escape piece overlaps with a point adjacent to the approaching ladder path and also adjacent to the escape piece.
partial ladder escape	A ladder escape that does not immediately connect to the edge, but forces a ladder that is closer to the edge.

path   Set of empty points and intermediate chains connecting two safe
                              groups.

pivot point   The terminal empty point of a step, from which further steps may
                              be taken. They are of structural importance and are the vulnerable
                              points of any path or step.

point   Hexagonal board cell, which may be *empty* or *occupied*.

safe group   Group within which every chain-to-chain pair is 0-connected.

safely connected   A connection that the opponent cannot block.

singleton group   Safe group comprised of a single piece or chain.

step   Step from a safe group composed of a *terminal point* and interme-
                              diate empty points.

swap option   An optional rule used to equalize Hex's strong first player advan-
                              tage: the player moving second may elect to swap colors instead
                              of playing on the second move of the game, effectively taking the
                              first player's move.

territory   That set of empty points touched by any of a player's pieces.

touching   A point (or piece) adjacent to another point (or piece).

unsafe group   Any group that is not safe (at least one chain-to-chain pair is not
                              0-connected).

valid board position   A board position that may have arisen during normal play.

I.2   Symbols

1..n   Move number for a specific piece.

a..z   Piece or chain.

(a, b...)   Set of pieces or chains.

C   Shorthand for a set of pieces or chains.

[i, j]   Rectangular coordinates (row, column).

A1..Z26   Alphanumeric notation for hexagonal board coordinates (letter = column, number = row).

$p..z$ Empty point.

$\underline{p}$ Pivot point.

●   Pivot or terminal point (diagrams only).

$\{p, q, r...\}$ Set of empty points which may contain pivot points.

$\{S\}$ Shorthand for a set of empty points.

I.3   Path Algebra

n-connected(a, b) Degree of connectivity between chains a and b: at least n moves are required to safely connect them.

n-group$(C\{S\})$ Group consisting of chain set C and empty point set S with worst case interchain connectivity n.

$<C\{S\}>$ Safe group consisting of chain set C and empty point set S.

$<a>$ Singleton group comprised of the chain a.

n-step$(<C\{S_1\}> \underline{p}\{S_2\})$ Step from safe group $<C\{S_1\}>$ and pivot point $\underline{p}$ of connectivity n and consisting of empty point set S_2.

n-path$(<C_1\{S_1\}> <C_2\{S_2\}> C_3\{S_3\})$

$\qquad$ Path between disjoint safe groups $<C_1\{S_1\}>$ and $<C_2\{S_2\}>$ of connectivity n and consisting of intermediate chain set C_3 and empty point set S_3.

$\oplus$ Path/step extension operator.

$\otimes$ Path/step consolidation operator.

I.4   Move Notation

31 H5   Thirty-first move of the game, played at point H5.

31 H5!   A good move.

31 H5!!   An excellent move (swings momentum in the player's favor or
                             sets up a win).

31 H5?   A questionable move.

31 H5??   A disastrously bad move (potential game loser).

1 F6 Swap   The second player has elected to play the swap option. They take
                             the opponent's color and effectively steal the first move.

Index

A

adjacent
  pieces 27, 53, 70, 97, 114
  points 25, 33
algorithmic board evaluation 127
  example 133
  features 145
  optimizations 144–145
  pseduocode 127

B

Berge's Hex problem 15, 240, 281
board 2, 327
  area, home and enemy 92–93
  even-sided 46, 152, 169
  formats 17, 19, 21
  labels 2, 18, 26
  Metahex 40, 316–317
  valid board position 5
    unique 5
bridge 29, 43, 70, 181, 227
  overlapping 91
  step 57, 97
Brouwer's Fixed-Point Theorem 15, 305

C

chain 1, 27, 59, 128
  winning 14, 31, 38
clarity 7
closing play 101, 203
  McCarthy Revenge rule 41
  resigning 51, 83, 186, 203, 212
  Surrounded Cell rule 38, 42, 313, 317
complexity 5, 15, 27, 77
connection 28, 43, 64, 84, 140
  loose 97, 159, 207
  safe 27, 42, 53, 65
coordinate system
  hexagonal 25–26
  rectangular 25
corner
  acute 33, 95, 153–154, 157, 305
  obtuse 32, 40, 95, 157

D

deadlock 33, 40
defense 49, 148, 228
  and attack 50, 182
  classic 46, 152, 157, 170, 227
  edge 99

determinism 4, 7, 51
   chance and luck 7, 205
diagonal
   long 152
   short 46, 93, 167
distance
   hexagonal 27, 30
   rectangular 27

E

empty point set 53, 56, 59, 69
   minimal 56, 72

F

fork. *See* move, fork

G

game
   abstract 4
   combinatorial 5, 15, 16, 196
   two-person, zero-sum, finite,
      deterministic, of strategy 4
game of
   Annular Hex. *See* Hex, variant,
      Cylindrical
   Backgammon 7, 152
   Bex. *See* Hex, variant, Beck's
   Birdcage. *See* game of Gale
   Black Path 323
   Black's Road Game. *See* game of
      Black Path
   Bridg-It. *See* game of Gale
   Chess 6, 8, 111
   Con-Tac-Tix. *See* Hex
   Connections. *See* game of Gale
   Fan 325
   Gale 7, 10, 319
   Go 3, 5, 8, 19
   Go-Moku 309
   Havannah 7, 317

Hexadame 325
Hexade 325
Hexbo 318
HexEmergo 325
Hexxagon 325
John. *See* Hex
Kaliko 324
Link 324
Macbeth 325
Metahex. *See* board, Metahex
Misere. *See* Hex, variant, Reverse
Mod Hex. *See* Hex, variant, Modulo
Nash. *See* Hex
Nine Men's Morris 6
Noughts and Crosses 309
Octiles 324
Othello 7
Pipeline 323
Poly-Y 315
Polygon. *See* Hex
Psyche Paths. *See* game of Kaliko
Rex. *See* Hex, variant, Reverse
Snakes Road. *See* game of Black Path
Square 33
Squiggly Road. *See* game of Black
   Path
Star 315
SuperStar 316
Tack 325
Tex. *See* Hex, variant, Infinite
Thoughtwave 322
Trax 322
Trinidad 324
TwixT 319
Ultima 7
Vortex. *See* Hex, variant, Voronoi
Winding Road. *See* game of Black Path
Y 314
   and Hex 14, 31, 96, 107, 111, 174,
      242
Gardner, Martin 3, 10, 13, 16, 241,
   305–306

goals 2, 5, 31, 44
  antipodean 312
graph 8
  cut/short 7–8
group 53, 55, 65, 129
  disjoint 56, 132
  importance 141
  singleton 53, 66, 129, 137

H

Hein, Piet 3, 13, 19, 36, 239, 303, 320
Hex
  classification 4, 5, 7, 16, 309
  history 3
  programs 4, 6, 8, 79, 146, 349, 351
  puzzles
    original 242
    previously published 239
    solutions 261
  rules 2
  sample games 287
    annotated 205
  variant
    Beck's 14, 154, 306, 310
    Cylindrical 16, 318, 325
    Handicap 282, 306, 311
    Infinite 313
    Kriegspiel 15, 310
    Map 36, 320
    Modulo 40, 312, 318, 325
    Non-Regular 16, 35, 320
    Reverse 14, 241, 310, 313
    Three-Dimensional 38, 325
    Three-Player 40
    Vex 310
    Voronoi 36, 321

K

killer heuristic 50
knot 21
  Reidemeister move 22

L

ladder 103, 130, 184, 231
  formation 105, 121
    bottleneck 105, 208, 210, 214
    plug 105, 121, 208, 213–214
ladder escape 106–107, 109, 184, 195, 232
  foil 112–113, 115, 117, 186, 234
    adjacent 114
    non-adjacent 116
  fork 111, 185, 189, 232, 266, 269, 272
  partial 120–121, 123
    cascading 123
    foldback 121
  piece 106, 108, 156, 166, 195, 201, 233
    embedded 107
  template 108, 113, 135, 185, 233
link 29, 44
  direct 70
  indirect 70
lookahead 196, 237
  long term 199, 237
  short term 197, 237

M

machine learning 7, 15
  neural network 7
  Temporal Difference 7
momentum 86, 90, 101, 204, 231
move
  blocking 41, 45, 112, 115, 121, 148, 157, 179, 187, 195, 227
    classic. *See* defense, classic
  foiling 50, 112–113, 115, 117, 164, 180, 186, 234
  forcing 87, 89, 91, 113, 173, 230
    forced reply 87, 103, 114, 150, 176, 200
  fork 29, 111, 113, 125, 173

killer 50, 86, 112, 178, 191
notation 205
red herring 190, 203
zwischenzug 173, 220

N

Nash, John 3, 14, 20, 305
neighborhood 25, 96
  immediate 25
  value 96, 108
    Worthless Triangle 96, 107

O

opening play 101, 225
  common plays 156
  first move 147, 154
    advantage 2, 14, 305
    dual challenges 3, 148
  first reply 148, 156
  swap option 2, 14, 84, 87, 152, 205,
    225
    when to swap 47, 153, 156
    where to open 101, 154

P

path 59, 132–133
  algebra 61
  analysis. *See* algorithmic board
    evaluation
  consolidation 63, 65, 133, 307
  convergent 178
  extension 61, 132
  spanning 64, 66, 83, 88, 101, 128,
    134, 203, 235, 306
point
  dual 29, 41, 63, 91, 149, 307
  edge 95
  interior 95

pivot 56, 62, 69, 85, 94, 133, 141,
    145, 148, 203
terminal 56, 69, 78, 130
vulnerable 56, 73, 76, 97, 114, 141,
    149, 161, 168, 179
polyhex 7, 16, 145, 347
polynomial space 15
  complete 15, 16
proof 303, 305
  acute corner opening loses 305
  existence 305
  one player must win 303
  strategy stealing 14–16, 305

S

Shannon, Claude 3, 8, 12, 13
Shannon Switching Game 8, 9, 11, 15,
    319
  on the edges 9
    and Gale 10, 12
  on the vertices 10
    and Hex 11
step 56, 60, 97, 129, 133
  consolidation 57, 131
  extension 58, 131
swap option. *See* opening play, swap
    option

T

template 69
  don't care point 72, 83
  edge 71, 82, 84, 100, 108, 130,
    134, 229
    4-row 73
    5-row 74
  interior 69, 82
  intrusion 74, 77, 88, 113, 144, 162,
    185, 190
    friendly 77

minimal 72, 77
multi-piece 78
territory 90, 101, 118, 120, 152, 154,
    178, 183, 201
    stealing 90, 102, 201, 283
threat 91, 93, 195, 228, 232, 235
    multiple 111, 173, 237
        Double Trouble 111, 174
    veiled 179

tiling
    planar 33
        irregular 36
        regular 35
        semi-regular 35
    three-dimensional 38, 325
        Buckminsterfullerene 38
touching. *See* adjacent